8th Mathematical Physics Meeting: Summer School and Conference on Modern Mathematical Physics 2014

AA005008

Belgrade, Serbia
24-31 August 2014

Editors:

B. Dragovich **I. Salom**

ISBN: 978-1-5108-2307-5

Printed from e-media with permission by:

Curran Associates, Inc.
57 Morehouse Lane
Red Hook, NY 12571

Some format issues inherent in the e-media version may also appear in this print version.

Printed by Curran Associates, Inc. (2016)

Belgrade Institute of Physics
Pregrevica 118
11080 Zemun – Belgrade
Serbia

Phone: +381 (0)11 37 13 000
Fax: +381 (0)11 31 62 190

info@ipb.ac.rs

Additional copies of this publication are available from:

Curran Associates, Inc.
57 Morehouse Lane
Red Hook, NY 12571 USA
Phone: 845-758-0400
Fax: 845-758-2633
Email: curran@proceedings.com
Web: www.proceedings.com

8th MATHEMATICAL PHYSICS MEETING: Summer School and Conference on Modern Mathematical Physics

Belgrade, August 24 - 31, 2014

Organizers

- Institute of Physics, Belgrade, Serbia
- Faculty of Mathematics, Belgrade, Serbia
- SEENET-MTP (Southeastern European Network in Mathematical and Theoretical Physics)

International Advisory Committee

International Organizing Committee

Loriano Bonora (Trieste, Italy)
Vladimir Dobrev (Sofia, Bulgaria)
Branko Dragovich (Belgrade, Serbia)
Viatcheslav Mukhanov (Munich, Germany)
Dejan Stojkovic (Buffalo, USA)
Igor Volovich (Moscow, Russia)
George Zoupanos (Athens, Greece)

Local Organizing Committee

Branko Dragovich, Co-Chairman (Inst. of Phys., Belgrade)
Igor Salom, Co-Chairman (Inst. of Phys., Belgrade)
Sanja Cirković (Institute of Physics, Belgrade)
Ljubica Davidović (Institute of Physics, Belgrade)
Goran Djordjević (Faculty of Sci. and Math., Niš)
Zoran Rakić (Faculty of Mathematics, Belgrade)

Sponsors

- **Ministry of Education, Science and Technological Development of the Republic of Serbia**
- **Southeastern European Network in Theoretical and Mathemat. Physics (SEENET-MTP Office Niš, Serbia)**
- **Project 174012: Geometry, Education and Visualization with Applications, Serbia**

PREFACE

This volume contains some reviews and original research contributions, which are presented at the **8th Mathematical Physics Meeting: Summer School and Conference on Modern Mathematical Physics**, held in Belgrade (Serbia), August 24–31, 2014 (http://www.mphys8.ipb.ac.rs). The programme of this meeting was mainly oriented towards some recent developments in gravity and cosmology, string and quantum field theory, and some relevant mathematical methods. We hope that articles presented here will be valuable literature not only for the participants of this meeting but also for many other PhD students and researchers in modern mathematical and theoretical physics. We are grateful to all authors for writing their contributions for these proceedings.

The previous seven meetings in this series of summer schools and conferences on modern mathematical physics were also held in Serbia: Sokobanja, 13–25 August 2001; Kopaonik, 1–12 September 2002; Zlatibor, 20–31 August 2004; Belgrade, 3–14 September 2006; Belgrade, 6–17 July 2008; Belgrade, 14–23 September 2010; and Belgrade, 9–19 September 2012. The corresponding proceedings of all these meetings were published by the Institute of Physics, Belgrade, and are available in the printed form as well as online at the meeting websites.

This eighth meeting took place at the "Milutin Milankovic Society" located in Belgrade downtown. There was conference dinner in Little Bay restaurant and excursion to Sremski Karlovci and Novi Sad. We hope that all attending this meeting will recall it as a useful and pleasant event, and will wish to participate again in the future.

We wish to thank all lecturers and other speakers for their interesting and valuable talks. We also thank all participants for their active participation. Financial support of our sponsors: *Ministry of Education, Science and Technological Development of the Republic of Serbia, Belgrade; ICTP – SEENET-MTP grant RRJ-09 "Cosmology and Strings", Niš, Serbia; Project 174012 (Geometry, Education and Visualization with Applications) Belgrade,* was very significant for realization of this activity.

June 2015

Editors

B. Dragovich

I. Salom

CONTENTS

Review and Research Works

Naturalness and string phenomenology in the LHC era

Ignatios Antoniadis*
Albert Einstein Center, Institute for Theoretical Physics
University of Bern, Sidlestrasse 5, CH-3012 Bern, Switzerland
LPTHE, UMR CNRS 7589, Université Pierre et Marie Curie, 75252 Paris
Ecole Polytechnique, F-91128 Palaiseau, France

ABSTRACT

I discuss the status of the mass hierarchy problem and prospects for beyond the Standard Model physics in the light of the Higgs scalar discovery at the LHC and the experimental searches for new physics. In particular, I discuss in this context low energy supersymmetry and large extra dimensions with low string scale.

1. Introduction

During the last few decades, physics beyond the Standard Model (SM) was guided from the problem of mass hierarchy. This can be formulated as the question of why gravity appears to us so weak compared to the other three known fundamental interactions corresponding to the electromagnetic, weak and strong nuclear forces. Indeed, gravitational interactions are suppressed by a very high energy scale, the Planck mass $M_P \sim 10^{19}$ GeV, associated to a length $l_P \sim 10^{-35}$ m, where they are expected to become important. In a quantum theory, the hierarchy implies a severe fine tuning of the fundamental parameters in more than 30 decimal places in order to keep the masses of elementary particles at their observed values. The reason is that quantum radiative corrections to all masses generated by the Higgs vacuum expectation value (VEV) are proportional to the ultraviolet cutoff which in the presence of gravity is fixed by the Planck mass. As a result, all masses are "attracted" to about 10^{16} times heavier than their observed values.

Besides compositeness, there are two main theories that have been proposed and studied extensively during the last years, corresponding to different approaches of dealing with the mass hierarchy problem. (1) Low energy supersymmetry with all superparticle masses in the TeV region. Indeed, in the limit of exact supersymmetry, quadratically divergent corrections to the Higgs self-energy are exactly cancelled, while in the softly broken case, they are cutoff by the supersymmetry breaking mass splittings. (2) TeV scale strings, in which quadratic divergences are cutoff by the string scale and

* e-mail address: antoniadis@itp.unibe.ch

low energy supersymmetry is not needed. Both ideas are experimentally testable at high-energy particle colliders and in particular at LHC.

On the other hand, the recent major discovery of the Higgs boson at the LHC with a mass around 126 GeV is so far compatible with the Standard Model within 2σ and its precision tests. It is also compatible with low energy supersymmetry, although with some degree of fine-tuning in its minimal version. Indeed, in the minimal supersymmetric Standard Model (MSSM), the lightest Higgs scalar mass m_h satisfies the following inequality:

$$m_h^2 \lesssim m_Z^2 \cos^2 2\beta + \frac{3}{(4\pi)^2}\frac{m_t^4}{v^2}\left[\ln\frac{m_{\tilde{t}}^2}{m_t^2} + \frac{A_t^2}{m_{\tilde{t}}^2}\left(1 - \frac{A_t^2}{12m_{\tilde{t}}^2}\right)\right] \lesssim (130GeV)^2 \,, \tag{1}$$

where the first term in the r.h.s. corresponds to the tree-level prediction and the second term includes the one loop corrections due to the top and stop loops. Here, $m_Z, m_t, m_{\tilde{t}}$ are the Z-boson, the top and stop quark masses, respectively, $v = \sqrt{v_1^2 + v_2^2}$ with v_i the VEVs of the two higgses, $\tan\beta = v_2/v_1$, and A_t the trilinear stop scalar coupling. Thus, a Higgs mass around 126 GeV requires a heavy stop $m_{\tilde{t}} \simeq 3$ TeV for vanishing A_t, or $A_t \simeq 3m_{\tilde{t}} \simeq 1.5$ TeV in the 'best' case. These values are obviously consistent with the present LHC bounds on supersymmetry searches, but they will certainly be probed in the next run at double energy. Theoretically, they imply a fine-tuning of the electroweak (EW) scale at the percent to per mille level. This fine-tuning can be alleviated in supersymmetric models beyond the MSSM, as discussed in the next session.

2. MSSM Higgs sector with dimension-five and dimension-six operators

Although extremely successful, the Standard Model or its supersymmetric version (MSSM) is not a fundamental theory, and this motivated the theoretical efforts to understand the nature of new physics beyond it. This search can be done using an effective field theory approach, in which the "new physics" is parametrised by effective operators. The power of this approach resides in arranging these operators in powers of $1/M_*$ where M_* is the scale of new physics that generated them. To improve the predictive power, one considers additional organising principles, such as: **(i)** symmetry constraints that these operators should respect, often inspired by phenomenology (for example: R-parity, lepton or baryon number conservation, etc). **(ii)** a truncation of the series of operators to a given order in the power of the inverse scale $1/M_*$. The effective low-energy Lagrangian then takes the form

$$\mathcal{L} = \mathcal{L}_0 + \sum_{i,n} \frac{c_{n,i}}{M_*^n}\,\mathcal{O}_{n,i} \tag{2}$$

where $\mathcal{L}_0$ is the SM or the MSSM Lagrangian; $\mathcal{O}_{n,i}$ is an operator of dimension $d = n + 4$ with the index i running over the set of operators of a given dimension; $c_{n,i}$ are some coefficients of order $\mathcal{O}(1)$. This description is appropriate for scales E which satisfy $E \ll M_*$. Constraints from phenomenology can then be used to set bounds on the scale of new physics M_*.

Regarding the origin of operators $\mathcal{O}_{n,i}$, they can be generated classically or at the quantum level. At the classical level, this can happen by integration of some new massive states, via the equations of motion and one then generates an infinite series. This can happen even in 4D renormalisable theories; indeed, even though the low energy interaction looks nonrenormalisable, it may actually point to a renormalisable theory valid up to a much higher scale (a familiar example is the Fermi interaction). Such effective operators are also generated at the quantum level, for example following compactification of a higher dimensional theory, by the radiative corrections associated with momentum and winding modes of the compactification [1, 2, 3, 4, 5].

The effects of these operators on the low energy observables can be comparable to the radiative effects of light states in the SM/MSSM [6] and this shows the importance of their study. In the following we shall study these effects to the case of the MSSM Higgs sector with additional operators of dimensions $d = 5$ and $d = 6$ [6, 7]. In particular we show that the mass of lightest SM-like Higgs can easily be increased close to the observed value by new physics in the region of few TeV. We then discuss the nature of the "new physics" behind the effective operators.

2.1. MSSM Higgs sector with d=5 and d=6 operators

In the leading order, new physics beyond the MSSM Higgs sector can manifest itself as operators of either $d = 5$ [6, 12, 13, 14] or $d = 6$ [7, 15] or both. If generated by the same new physics, by comparing $\mathcal{O}(1/M^*)$ and $\mathcal{O}(1/M_*^2)$ terms one can estimate when the series expansion in $1/M_*$ breaks down. There is only one operator in the Higgs sector of dimension $d = 5$:

$$\mathcal{L}_1 = \frac{1}{M_*} \int d^2\theta \; \lambda'_H(S) \, (H_2.H_1)^2 + h.c. \;\; = \;\; 2\,\zeta_{10}\,(h_2.h_1)(h_2.F_1 + F_2.h_1) + \zeta_{11}\, m_0\,(h_2.h_1)^2 + h.c, \qquad (3)$$

where $\lambda'_H(S)/M_* = \zeta_{10} + \zeta_{11}\, m_0\, \theta\theta$ and $\zeta_{10}, \zeta_{11} \sim 1/M_*$. It can be generated by integrating out a massive gauge singlet or $SU(2)$ triplet. Indeed, in the MSSM with a massive gauge singlet, with an F-term of type $M_*\Sigma^2 + \Sigma\, H_1.H_2$, when integrating out Σ generates $\mathcal{L}_1$. With the standard notation, here $S = m_0\theta^2$ is the spurion superfield and m_0 the supersymmetry breaking scale.

We assume that $m_0 \ll M_*$, so that the effective approach is reliable. If this is not respected and the "new physics" is represented by "light" states (like the MSSM states), the $1/M_*$ expansion is not reliable and one should work in a setup where these are not integrated out.

The list of $d = 6$ operators is longer [16]:

$$\mathcal{O}_1 = \frac{1}{M_*^2} \int d^4\theta \; \mathcal{Z}_1 \; (H_1^\dagger \, e^{V_1} \, H_1)^2,$$

$$
\begin{aligned}
\mathcal{O}_2 &= \frac{1}{M_*^2}\int d^4\theta\; \mathcal{Z}_2\; (H_2^\dagger\, e^{V_2}\, H_2)^2, \\
\mathcal{O}_3 &= \frac{1}{M_*^2}\int d^4\theta\; \mathcal{Z}_3\; (H_1^\dagger\, e^{V_1}\, H_1)\,(H_2^\dagger\, e^{V_2}\, H_2), \\
\mathcal{O}_4 &= \frac{1}{M_*^2}\int d^4\theta\; \mathcal{Z}_4\; (H_2.\, H_1)\,(H_2.\, H_1)^\dagger, \\
\mathcal{O}_5 &= \frac{1}{M_*^2}\int d^4\theta\; \mathcal{Z}_5\; (H_1^\dagger\, e^{V_1}\, H_1)\; H_2.\, H_1 + h.c. \\
\mathcal{O}_6 &= \frac{1}{M_*^2}\int d^4\theta\; \mathcal{Z}_6\; (H_2^\dagger\, e^{V_2}\, H_2)\; H_2.\, H_1 + h.c. \\
\mathcal{O}_7 &= \frac{1}{M_*^2}\int d^2\theta\; \mathcal{Z}_7\, \mathrm{Tr}\, W^\alpha\, W_\alpha\, (H_2\, H_1) + h.c. \\
\mathcal{O}_8 &= \frac{1}{M_*^2}\int d^4\theta\; \mathcal{Z}_8\; (H_2\, H_1)^2 + h.c.
\end{aligned} \tag{4}
$$

where $W^\alpha = (-1/4)\,\overline{D}^2 e^{-V} D^\alpha\, e^V$ is the chiral field strength of $SU(2)_L$ or $U(1)_Y$ vector superfields V_w and V_Y respectively. Also $V_{1,2} = V_w^a(\sigma^a/2) + (\mp 1/2)\, V_Y$ with the upper (minus) sign for V_1. Finally, the wavefunction coefficients are spurion dependent and have the structure

$$
(1/M_*^2)\, \mathcal{Z}_i(S, S^\dagger) = \alpha_{i0} + \alpha_{i1}\, m_0\, \theta\theta + \alpha_{i1}^*\, m_0\, \overline{\theta\theta} + \alpha_{i2}\, m_0^2\, \theta\theta\overline{\theta\theta}, \qquad \alpha_{ij} \sim 1/M_*^2. \tag{5}
$$

Regarding the origin of these operators: $\mathcal{O}_{1,2,3}$ can be generated in the MSSM by an additional, massive $U(1)'$ gauge boson or $SU(2)$ triplets, when integrated out [12]. $\mathcal{O}_4$ can be generated by a massive gauge singlet or $SU(2)$ triplet, while $\mathcal{O}_{5,6}$ can be generated by a combination of $SU(2)$ doublets and massive gauge singlet. $\mathcal{O}_7$ is essentially a threshold correction to the gauge coupling, with a moduli field replaced by the Higgs. $\mathcal{O}_8$ exists only in non-susy case, but is generated when removing a $d = 5$ derivative operator by field redefinitions [6], so we keep it.

2.2. Higgs mass corrections from $d = 5$ and $d = 6$ operators.

With the above set of independent, effective operators, one finds the scalar potential V and its EW minimum which is perturbed by $\mathcal{O}(1/M_*^2)$ corrections from that of the MSSM. From V one computes the mass of CP-odd/even Higgs fields:

$$
m_A^2 = (m_A^2)_{\rm MSSM} - \frac{2\,\zeta_{10}\,\mu_0\, v^2}{\sin 2\beta} + 2\, m_0\, \zeta_{11}\, v^2 + \delta m_A^2, \quad \delta m_A^2 = \mathcal{O}(1/M_*^2) \tag{6}
$$

for the pseudoscalar Higgs, with $(m_A^2)_{\rm MSSM}$ the MSSM value, with δm_A^2 due to $\mathcal{O}(1/M_*^2)$ corrections from $d = 5$ and $d = 6$ operators. For the

CP-even Higgs one has [6, 12, 14]

$$m_{h,H}^2 = (m_{h,H}^2)_{\text{MSSM}} + (2\,\zeta_{10}\,\mu_0)\,v^2\,\sin 2\beta\left[1 \pm \frac{m_A^2+m_Z^2}{\sqrt{\tilde w}}\right] + \frac{(-2\,\zeta_{11}\,m_0)\,v^2}{2}\left[1 \mp \frac{(m_A^2-m_Z^2)\,\cos^2 2\beta}{\sqrt{\tilde w}}\right] + \delta m_{h,H}^2, \quad \text{where} \quad \delta m_{h,H}^2 = \mathcal{O}(1/M_*^2) \tag{7}$$

The upper (lower) signs correspond to h (H), and $\tilde w \equiv (m_A^2+m_Z^2)^2 - 4\,m_A^2\,m_Z^2\,\cos^2 2\beta$. With this result one can show that the mass m_h can be increased near the observed value, also with the help of quantum corrections [6, 12, 13, 14].

Regarding the $\mathcal{O}(1/M_*^2)$ corrections of $\delta m_{h,H}^2, \delta m_A^2$ and $\delta m_{h,H}^2$ of (6), (7), in the general case of including all operators and their associated supersymmetry breaking, they have a complicated form. Exact expressions can be found in [7, 15]. For most purposes, an expansion of these in $1/\tan\beta$ is accurate enough. At large $\tan\beta$, $d = 6$ operators bring corrections comparable to those of $d = 5$ operators. The relative $\tan\beta$ enhancement of $\mathcal{O}(1/M_*^2)$ corrections compensates for the extra suppression that these have relative to $\mathcal{O}(1/M_*)$ operators (which involve both h_1, h_2 and are not enhanced in this limit). Note however that in some models only $d = 6$ operators may be present, depending on the details of the "new physics" generating the effective operators.

Let us present the correction $\mathcal{O}(1/M^2)$ to m_h^2 for the case m_A is kept fixed to an appropriate value. The result is, assuming $m_A > m_Z$, (otherwise δm_h^2 and δm_H^2 are exchanged):

$$\begin{aligned}\delta m_h^2 = & - 2\,v^2\left[\alpha_{22}\,m_0^2 + (\alpha_{30}+\alpha_{40})\mu_0^2 + 2\alpha_{61}\,m_0\,\mu_0 - \alpha_{20}\,m_Z^2\right] \\ & - (2\,\zeta_{10}\,\mu_0)^2\,v^4\,(m_A^2-m_Z^2)^{-1} + v^2\cot\beta\Big[(m_A^2-m_Z^2)^{-1}\Big(4\,m_A^2 \\ & \big(\,(2\alpha_{21}+\alpha_{31}+\alpha_{41}+2\alpha_{81})\,m_0\,\mu_0 + (2\alpha_{50}+\alpha_{60})\,\mu_0^2 + \alpha_{62}\,m_0^2\big) \\ & -(2\alpha_{60}-3\alpha_{70})\,m_A^2\,m_Z^2 - (2\alpha_{60}+\alpha_{70})\,m_Z^4\Big) \\ & + 8\,(m_A^2+m_Z^2)\,(\mu_0\,m_0\,\zeta_{10}\,\zeta_{11})\,v^2/(m_A^2-m_Z^2)^2\Big] + \mathcal{O}(1/\tan^2\beta)\end{aligned} \tag{8}$$

The mass corrections in (8) must be added to the rhs of eq.(7) to obtain the full value of m_h^2. Together with (4), (5), these corrections identify the operators of $d = 6$ with the largest contributions, which is important for model building beyond the MSSM Higgs sector. These operators are $\mathcal{O}_{2,3,4}$ in the absence of supersymmetry breaking and $\mathcal{O}_{2,6}$ when this is broken. It is preferable, however, to increase m_h^2 by supersymmetric rather than supersymmetry-breaking effects of the effective operators, because the latter are less under control in the effective approach; also, one would favour a supersymmetric solution to the fine-tuning problem associated with increasing the MSSM Higgs mass. Therefore $\mathcal{O}_{2,3,4}$ are the leading operators, with the remark that $\mathcal{O}_2$ has a smaller effect, of order $(m_Z/\mu_0)^2$ relative to $\mathcal{O}_{3,4}$ (for similar α_{j0}, $j=2,3,4$). At smaller $\tan\beta$, $\mathcal{O}_{5,6}$ can also give significant contributions, while $\mathcal{O}_7$ has a relative suppression factor $(m_Z/\mu_0)^2$. Note that we kept all operators $\mathcal{O}_i$ independent. By doing so, one can easily single out the individual contribution of each operator, which helps in model building, since not all operators are present in a specific model.

One limit to consider is that where the operators of $d=6$ have coefficients such that their contributions add up to maximise δm_h^2. Since α_{ij} are not known, one can choose: $-\alpha_{22} = -\alpha_{61} = -\alpha_{30} = -\alpha_{40} = \alpha_{20} > 0$. In this case, at large $\tan\beta$:

$$\delta m_h^2 \approx 2\,v^2\alpha_{20}[m_0^2 + 2\,m_0\mu_0 + 2\,\mu_0^2 + m_Z^2] \tag{9}$$

A simple numerical example is illustrative. For $m_0 = 1$ TeV, $\mu_0 = 350$ GeV, and with $v \approx 246$ GeV, one has $\delta m_h^2 \approx 2.36\,\alpha_{20}\,\times 10^{11}$ (GeV)^2. Assuming $M_* = 10$ TeV and ignoring $d=5$ operators, with $\alpha_{20} \sim 1/M_*^2$ and the MSSM value of m_h taken to be its upper classical limit m_Z (reached for large $\tan\beta$), we obtain an increase of m_h from $d=6$ operators alone of about $\Delta m_h = 12.15$ GeV to $m_h \approx 103$ GeV. An increase of α_{20} by a factor of 2.5 to $\alpha_{20} \sim 2.5/M_*^2$ would give $\Delta m_h \approx 28$ GeV to $m_h \approx 119.2$ GeV, which is already above the LEP bound. Note that this increase is realised even for a scale M_* of new physics beyond the LHC reach.

The above choice of $M_* = 10$ TeV was partly motivated by the fine-tuning results [13] and on convergence grounds: the expansion parameter of our effective analysis is m_q/M_* where m_q is any scale of the theory, in particular it can be m_0. For a susy breaking scale $m_0 \sim \mathcal{O}(1)$ TeV (say $m_0 = 3$ TeV) and $c_{1,2}$ or α_{ij} of $\mathcal{Z}_i(S, S^\dagger)$ of order unity (say $c_{1,2} = 2.5$) one has for $M_* = 10$ TeV that $c_{1,2}\,m_0/M_* = 0.75$ which is already close to unity, and at the limit of validity of the effective expansion in powers of $1/M_*$. To conclude, even for a scale of "new physics" above the LHC reach, one can still classically increase m_h to near the LHC measured value.

3. Strings and extra dimensions

The appropriate and most convenient framework for low energy supersymmetry and grand unification is the perturbative heterotic string. Indeed, in this theory, gravity and gauge interactions have the same origin, as massless modes of the closed heterotic string, and they are unified at the string scale M_s. As a result, the Planck mass is predicted to be proportional to M_s:

$$M_P = M_s/g\,, \tag{10}$$

where g is the gauge coupling. In the simplest constructions all gauge couplings are the same at the string scale, given by the four-dimensional (4d) string coupling, and thus no grand unified group is needed for unification.

In our conventions $\alpha_{\rm GUT} = g^2 \simeq 0.04$, leading to a discrepancy between the string and grand unification scale $M_{\rm GUT}$ by almost two orders of magnitude. Explaining this gap introduces in general new parameters or a new scale, and the predictive power is essentially lost. This is the main defect of this framework, which remains though an open and interesting possibility.

The other other perturbative framework that has been studied extensively in the more recent years is type I string theory with D-branes. Unlike in the heterotic string, gauge and gravitational interactions have now different origin. The latter are described again by closed strings, while the former emerge as excitations of open strings with endpoints confined on D-branes [17]. This leads to a braneworld description of our universe, which should be localized on a hypersurface, i.e. a membrane extended in p spatial dimensions, called p-brane (see Fig. 1). Closed strings propagate in all nine dimensions of string theory: in those extended along the p-brane, called parallel, as well as in the transverse ones. On the contrary, open strings are attached on the p-brane. Obviously, our p-brane world must have at least

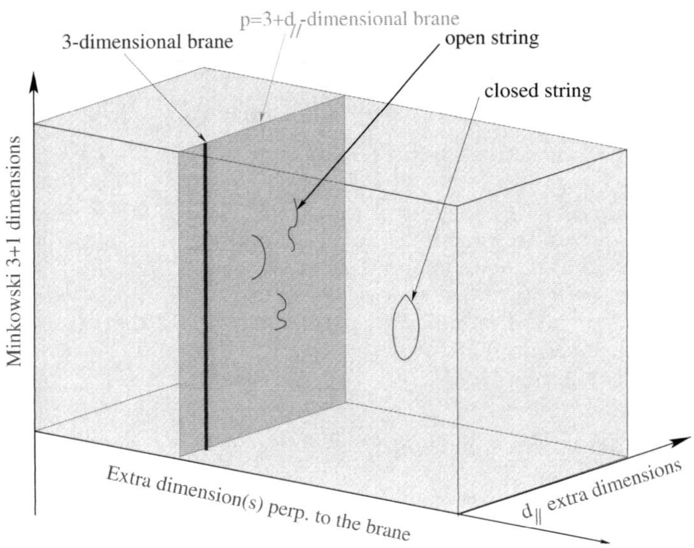

Figure 1: D-brane world universe in type I string framework.

the three known dimensions of space. But it may contain more: the extra $d_{\|} = p - 3$ parallel dimensions must have a finite size, in order to be unobservable at present energies, and can be as large as $\text{TeV}^{-1} \sim 10^{-18}$ m [18]. On the other hand, transverse dimensions interact with us only gravitationally and experimental bounds are much weaker: their size should be less than about 0.1 mm [19]. In the following, I review the main properties and experimental signatures of low string scale models [20].

3.1. Framework of low scale strings

In type I theory, the different origin of gauge and gravitational interactions implies that the relation between the Planck and string scales is not linear as (10) of the heterotic string. The requirement that string theory should be weakly coupled, constrain the size of all parallel dimensions to be of order of the string length, while transverse dimensions remain unrestricted. Assuming an isotropic transverse space of $n = 9 - p$ compact dimensions of common radius $R_\perp$, one finds:

$$M_P^2 = \frac{1}{g_s^2} M_s^{2+n} R_\perp^n \,, \qquad g_s \simeq g^2 \,. \tag{11}$$

where g_s is the string coupling. It follows that the type I string scale can be chosen hierarchically smaller than the Planck mass [21, 20] at the expense of introducing extra large transverse dimensions felt only by gravity, while keeping the string coupling small [20]. The weakness of 4d gravity compared to gauge interactions (ratio M_W/M_P) is then attributed to the largeness of the transverse space $R_\perp$ compared to the string length $l_s = M_s^{-1}$.

An important property of these models is that gravity becomes effectively $(4+n)$-dimensional with a strength comparable to those of gauge interactions at the string scale. The first relation of Eq. (11) can be understood as a consequence of the $(4+n)$-dimensional Gauss law for gravity, with $M_*^{(4+n)} = M_s^{2+n}/g^4$ the effective scale of gravity in $4+n$ dimensions. Taking $M_s \simeq 1$ TeV, one finds a size for the extra dimensions $R_\perp$ varying from 10^8 km, .1 mm, down to a Fermi for $n = 1, 2$, or 6 large dimensions, respectively. This shows that while $n = 1$ is excluded, $n \geq 2$ is allowed by present experimental bounds on gravitational forces [19, 22]. Thus, in these models, gravity appears to us very weak at macroscopic scales because its intensity is spread in the "hidden" extra dimensions. At distances shorter than $R_\perp$, it should deviate from Newton's law, which may be possible to explore in laboratory experiments.

3.2. Experimental implications in accelerators

We now turn to the experimental predictions of TeV scale strings. Their main implications in particle accelerators are of four types, in correspondence with the four different sectors that are generally present:

1. New compactified parallel dimensions; In this case $RM_s \gtrsim 1$, and the associated compactification scale $R_\parallel^{-1}$ would be the first scale of new physics that should be found increasing the beam energy [18, 23]. The main consequence is the existence of KK excitations for all SM particles that propagate along the extra parallel dimensions. These can be produced on-shell at LHC as new resonances [24].
2. New extra large transverse dimensions and low scale quantum gravity,. The main experimental signal is gravitational radiation in the bulk from any physical process on the world-brane [25].

3. Genuine string and quantum gravity effects. Direct production of string resonances in hadron colliders leads generically to a universal deviation from Standard Model in jet distribution [26]. In particular, the first Regge excitation of the gluon has spin 2 and a width an order of magnitude lower than the string scale, leading to a characteristic peak in dijet production; similarly, the first excitations of quarks have spin 3/2.
4. Extra $U(1)$'s arising generically in D-brane models as part of unitary gauge group factors. They obtain in general masses due to four- or higher-dimensional anomalies, via the so-called Green-Schwarz anomaly cancellation mechanism involving axionic fields from the closed string sector. The resulting masses are therefore suppressed by a loop factor compared to the string scale. From the low energy point of view, they gauge global symmetries of the Standard Model, such as the baryon and lepton number. An important property of the anomaly cancellation mechanism is that the anomalous $U(1)$ gauge bosons acquire masses leaving behind the corresponding global symmetries unbroken in perturbation theory. Thus, this is a way to guarantee proton stability (from unbroken baryon number) and avoid large Majorana neutrino masses (from unbroken lepton number) due to dimension-5 operators involving two higgses and two leptons that are suppressed only by the TeV string scale. Such extra $U(1)$s have interesting properties and distinct experimental signatures [27, 28, 29].
5. Concerning possible micro-black hole production, note that a string size black hole has a horizon radius $r_H \sim 1$ in string units, while the Newton's constant behaves as $G_N \sim g_s^2$. It follows that the mass of a d-dimensional black hole is [30]: $M_{\rm BH} \sim r_H^{d/2-1}/G_N \simeq 1/g_s^2$. Using the value of the SM gauge couplings $g_s \simeq g^2 \sim 0.1$, one finds that the energy threshold $M_{\rm BH}$ of micro-black hole production is about four orders of magnitude higher than the string scale, implying that one would produce 10^4 string states before reaching $M_{\rm BH}$.

3.3. Electroweak symmetry breaking

Non-supersymmetric TeV strings offer also a framework to realize gauge symmetry breaking radiatively. Indeed, from the effective field theory point of view, one expects quadratically divergent one-loop contributions to the masses of scalar fields. The divergences are cut off by M_s and if the corrections are negative, they can induce electroweak symmetry breaking and explain the mild hierarchy between the weak and a string scale at a few TeV, in terms of a loop factor [31]. More precisely, in the minimal case of one Higgs doublet H, the scalar potential is:

$$V = \lambda(H^\dagger H)^2 + \mu^2(H^\dagger H)\,, \tag{12}$$

where λ arises at tree-level. Moreover, in any model where the Higgs field comes from an open string with both ends fixed on the same brane stack, it is given by an appropriate truncation of a supersymmetric theory. On the other hand, μ^2 is generated at one loop:

$$\mu^2 = -\varepsilon^2 g^2 M_s^2 \,, \tag{13}$$

where ε is a loop factor that can be estimated from a toy model computation and varies in the region $\epsilon \sim 10^{-1} - 10^{-3}$.

Indeed, consider for illustration a simple case where the whole one-loop effective potential of a scalar field can be computed. We assume for instance one extra dimension compactified on a circle of radius $R > 1$ (in string units). An interesting situation is provided by a class of models where a non-vanishing VEV for a scalar (Higgs) field ϕ results in shifting the mass of each KK excitation by a constant $a(\phi)$: $M_m^2 = [(m + a(\phi)/R]^2$, with m the KK integer momentum number. Such mass shifts arise for instance in the presence of a Wilson line, $a = q \oint \frac{dy}{2\pi} A$, where A is the internal component of a gauge field and q the charge of a given state under the corresponding generator. A straightforward computation shows that the ϕ-dependent part of the one-loop effective potential is given by [32]:

$$V_{eff} = -Tr(-)^F \frac{R}{32\,\pi^{3/2}} \sum_n e^{2\pi i n a} \int_0^\infty dl\; l^{3/2} f_s(l)\; e^{-\pi^2 n^2 R^2 l} \tag{14}$$

where $F = 0, 1$ for bosons and fermions, respectively. We have included a regulating function $f_s(l)$ which contains for example the effects of string oscillators. To understand its role we will consider the two limits $R >> 1$ and $R << 1$. In the first case only the $l \to 0$ region contributes to the integral. This means that the effective potential receives sizable contributions only from the infrared (field theory) degrees of freedom. In this limit we would have $f_s(l) \to 1$. For example, in the string model considered in [31]: $f_s(l) = \left[\frac{1}{4l}\frac{\theta_2}{\eta^3}(il + \frac{1}{2})\right]^4 \to 1$ for $l \to 0$, and the field theory result is finite and can be explicitly computed. As a result of the Taylor expansion around $a = 0$, we are able to extract the one-loop contribution to the coefficient of the term of the potential quadratic in the Higgs field. It is given by a loop factor times the compactification scale [32]. One thus obtains $\mu^2 \sim g^2/R^2$ up to a proportionality constant which is calculable in the effective field theory. On the other hand, if we consider $R \to 0$, which by T-duality corresponds to taking the extra dimension as transverse and very large, the one-loop effective potential receives contributions from the whole tower of string oscillators as appearing in $f_s(l)$, leading to squared masses given by a loop factor times M_s^2, according to eq. (13).

More precisely, from the expression (14), one finds:

$$\varepsilon^2(R) = \frac{1}{2\pi^2} \int_0^\infty \frac{dl}{(2\,l)^{5/2}} \frac{\theta_2^4}{4\eta^{12}} \left(il + \frac{1}{2}\right) R^3 \sum_n n^2 e^{-2\pi n^2 R^2 l} \,. \tag{15}$$

which is plotted in Fig. 2. For the asymptotic value $R \to 0$ (corresponding

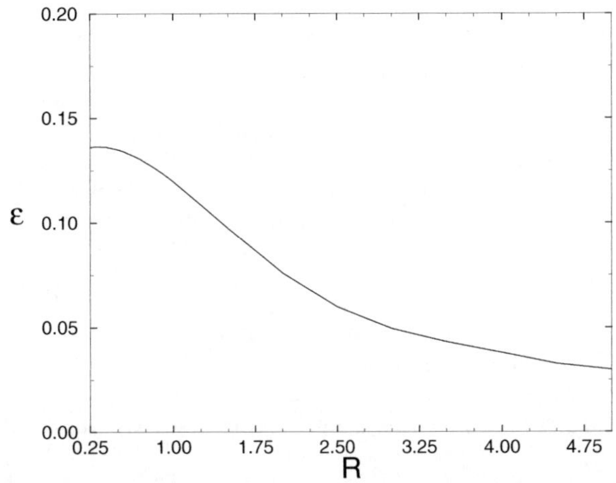

Figure 2: The coefficient ε of the one loop Higgs mass (13).

upon T-duality to a large transverse dimension of radius $1/R$), $\varepsilon(0) \simeq 0.14$, and the effective cut-off for the mass term is M_s, as can be seen from Eq. (13). At large R, $\mu^2(R)$ falls off as $1/R^2$, which is the effective cut-off in the limit $R \to \infty$, as we argued above, in agreement with field theory results in the presence of a compactified extra dimension.In fact, in the limit $R \to \infty$, an analytic approximation to $\varepsilon(R)$ gives:

$$\varepsilon(R) \simeq \frac{\varepsilon_\infty}{M_s\, R}, \qquad \varepsilon_\infty^2 = \frac{3\,\zeta(5)}{4\,\pi^4} \simeq 0.008\,. \tag{16}$$

The potential (12) has the usual minimum, given by the VEV of the neutral component of the Higgs doublet $v = \sqrt{-\mu^2/\lambda}$. Furthermore, from (13), one can compute M_s in terms of the Higgs mass $m_H^2 = -2\mu^2$: $M_s = \frac{m_H}{\sqrt{2}\,g\varepsilon}$, yielding naturally values in the TeV range.

References

[1] D. M. Ghilencea and H. M. Lee, JHEP **0509** (2005) 024.

[2] D. M. Ghilencea, Phys. Rev. D **70** (2004) 045018.

[3] S. Groot Nibbelink, M. Hillenbach, Phys. Lett. B **616** (2005) 125.

[4] S. Groot Nibbelink, M. Hillenbach, Nucl.Phys. B **748** (2006) 60.

[5] J. F. Oliver, J. Papavassiliou and A. Santamaria, Phys. Rev. D **67** (2003) 125004.

[6] I. Antoniadis, E. Dudas, D. Ghilencea, P. Tziveloglou, Nucl. Phys. B **808** (2009) 155.

[7] I. Antoniadis, E. Dudas, D. Ghilencea, P.Tziveloglou, Nucl. Phys. B **831** (2010) 133

[8] G. F. Giudice, C. Grojean, A. Pomarol and R. Rattazzi, JHEP **0706** (2007) 045.

[9] S. Cassel, D. M. Ghilencea, G. G. Ross, Nucl. Phys. B **835** (2010) 110.

[10] S. Cassel, D. M. Ghilencea and G. G. Ross, Phys. Lett. B **687** (2010) 214.

[11] R. Barate *et al.* [LEP Working Group for Higgs boson searches and ALEPH, DELPHI, L3, OPAL Collaborations, Phys. Lett. B **565** (2003) 61; S. Schael *et al.* ALEPH, DELPHI, L3, OPAL and LEP Working Group for Higgs Boson Searches, Eur. Phys. J. C**47**(2006)547

[12] M. Dine, N. Seiberg, S. Thomas, Phys. Rev. D **76** (2007) 095004.

[13] S. Cassel, D. Ghilencea, G. G. Ross, Nucl. Phys. B **825** (2010) 203.

[14] K. Blum, C. Delaunay and Y. Hochberg, arXiv:0905.1701 [hep-ph].

[15] M. Carena, K. Kong, E. Ponton and J. Zurita, arXiv:0909.5434 [hep-ph]. M. Carena, E. Ponton and J. Zurita, arXiv:1005.4887 [hep-ph].

[16] D. Piriz and J. Wudka, Phys. Rev. D **56** (1997) 4170. A. Brignole, J. A. Casas, J. R. Espinosa and I. Navarro,

[17] C. Angelantonj and A. Sagnotti, *Phys. Rept.* **371** (2002) 1 [Erratum-ibid. **376** (2003) 339] [arXiv:hep-th/0204089].

[18] I. Antoniadis, *Phys. Lett.* B **246** (1990) 377.

[19] D. J. Kapner, T. S. Cook, E. G. Adelberger, J. H. Gundlach, B. R. Heckel, C. D. Hoyle and H. E. Swanson, Phys. Rev. Lett. **98** (2007) 021101.

[20] N. Arkani-Hamed, S. Dimopoulos and G. R. Dvali, *Phys. Lett.* B **429** (1998) 263 [arXiv:hep-ph/9803315]; I. Antoniadis, N. Arkani-Hamed, S. Dimopoulos and G. R. Dvali, *Phys. Lett.* B **436** (1998) 257 [arXiv:hep-ph/9804398].

[21] J. D. Lykken, *Phys. Rev.* D **54** (1996) 3693 [arXiv:hep-th/9603133].

[22] J.C. Long and J.C. Price, *Comptes Rendus Physique* **4** (2003) 337; R.S. Decca, D. Lopez, H.B. Chan, E. Fischbach, D.E. Krause and C.R. Jamell, *Phys. Rev. Lett.* **94** (2005) 240401; R.S. Decca et al., arXiv:0706.3283 [hep-ph]; S.J. Smullin, A.A. Geraci, D.M. Weld, J. Chiaverini, S. Holmes and A. Kapitulnik, arXiv:hep-ph/0508204; H. Abele, S. Haeßler and A. Westphal, in 271th WE-Heraeus-Seminar, Bad Honnef (2002).

[23] I. Antoniadis and K. Benakli, *Phys. Lett.* B **326** (1994) 69.

[24] I. Antoniadis, K. Benakli and M. Quirós, *Phys. Lett.* B **331** (1994) 313 and *Phys. Lett.* B **460** (1999) 176; P. Nath, Y. Yamada and M. Yamaguchi, *Phys. Lett.* B **466** (1999) 100 T. G. Rizzo and J. D. Wells, *Phys. Rev.* D **61** (2000) 016007; T. G. Rizzo, *Phys. Rev.* D **61** (2000) 055005; A. De Rujula, A. Donini, M. B. Gavela and S. Rigolin, *Phys. Lett.* B **482** (2000) 195.

[25] G.F. Giudice, R. Rattazzi and J.D. Wells, *Nucl. Phys.* B **544** (1999) 3; E.A. Mirabelli, M. Perelstein and M.E. Peskin, *Phys. Rev. Lett.* **82** (1999) 2236; T. Han, J.D. Lykken and R. Zhang, *Phys. Rev.* D **59** (1999) 105006; K. Cheung, W.-Y. Keung, *Phys. Rev.* D **60** (1999) 112003; C. Balázs *et al.*, *Phys. Rev. Lett.* **83** (1999) 2112; J.L. Hewett, *Phys. Rev. Lett.* **82** (1999) 4765.

[26] L.A. Anchordoqui, H. Goldberg, D. Lust, S. Nawata, S. Stieberger and T.R. Taylor, Phys. Rev. Lett. **101** (2008) 241803 [arXiv:0808.0497 [hep-ph]].

[27] I. Antoniadis, E. Kiritsis and T. N. Tomaras, Phys. Lett. B **486**, 186 (2000) [arXiv:hep-ph/0004214]; I. Antoniadis, E. Kiritsis and J. Rizos, Nucl. Phys. B **637**, 92 (2002) [arXiv:hep-th/0204153]; I. Antoniadis, E. Kiritsis, J. Rizos and T. N. Tomaras, Nucl. Phys. B **660**, 81 (2003) [arXiv:hep-th/0210263].

[28] G. Shiu and S.-H. H. Tye, *Phys. Rev.* D **58** (1998) 106007; Z. Kakushadze and S.-H. H. Tye, *Nucl. Phys.* B **548** (1999) 180; L. E. Ibáñez, C. Muñoz and S. Rigolin, *Nucl. Phys.* B **553** (1999) 43.

[29] L. A. Anchordoqui, I. Antoniadis, H. Goldberg, X. Huang, D. Lust and T. R. Taylor, Phys. Rev. D **85** (2012) 086003 [arXiv:1107.4309 [hep-ph]].

[30] G.T. Horowitz and J. Polchinski, Phys. Rev. D **55** (1997) 6189.

[31] I. Antoniadis, K. Benakli and M. Quirós, *Nucl. Phys.* B **583** (2000) 35.

[32] I. Antoniadis, K. Benakli and M. Quiros, *New Jour. Phys.* **3** (2001) 20.

Orbits of S2 star in Yukawa gravity: simulations vs observations *

D. Borka[†]
Atomic Physics Laboratory (040), Vinča Institute of Nuclear Sciences
University of Belgrade, P.O. Box 522, 11001 Belgrade, Serbia
P. Jovanović[‡]
Astronomical Observatory, Volgina 7, 11060 Belgrade, Serbia
V. Borka Jovanović[§]
Atomic Physics Laboratory (040), Vinča Institute of Nuclear Sciences
University of Belgrade, P.O. Box 522, 11001 Belgrade, Serbia
A. F. Zakharov[¶]
Institute of Theoretical and Experimental Physics, B. Cheremushkinskaya 25, 117259 Moscow, Russia
Bogoliubov Laboratory for Theoretical Physics, JINR, 141980 Dubna, Russia

Abstract

We investigate some possible observational signatures of Yukawa gravity at Galactic scales. We performed simulations in Yukawa gravity potential and analyzed the obtained trajectories of S2 star around Galactic Centre. We compare the obtained theoretical results for S2 star orbit in the Yukawa potential with two independent sets of observations of the S2 star obtained by New Technology Telescope/Very Large Telescope (NTT/VLT), as well as by Keck telescope. Our results show that the most probable value for the parameter Λ in Yukawa gravity potential in the case of S2 star is around 5000 - 7000 AU. At the same time, we were not able to obtain reliable constrains on the universal constant δ of Yukawa gravity, because the current observations of S2 star indicated that it may be highly correlated with parameter Λ. We obtained that the Yukawa gravity potential induces precession of S2 star orbit in the same direction as General Relativity for $\delta > 0$ and for $\delta < -1$, and in the opposite direction for $-1 < \delta < 0$.

pacs 04.50.Kd, 04.80.Cc, 04.25.Nx, 04.50.-h

* Work supported by the Ministry of Education, Science and Technological Development of the Republic of Serbia through the project 176003 "Gravitation and the large scale structure of the Universe".

† e-mail address: dusborka@vinca.rs
‡ e-mail address: pjovanovic@aob.rs
§ e-mail address: vborka@vinca.rs
¶ e-mail address: zakharov@itep.ru

1. Introduction

Extended Theories of Gravity [1] have been proposed like alternative approaches to Newtonian gravity in order to explain galactic and extragalactic dynamics without introducing dark matter [2, 3]. $f(R)$ gravity is a type of modified gravity which generalizes Einstein's General Relativity and it was first proposed in 1970 by Buchdahl [4]. $f(R)$ gravity presented a family of models, each one defined by a different function of the Ricci scalar. The simplest case is just the General Relativity. In the case of $f(R)$ gravity, one assumes a generic function f of the Ricci scalar R (in particular, analytic functions) and searches for a theory of gravity having suitable behavior at different scales: for short distances, Solar system, spiral galaxies and galaxy clusters.

Yukawa-like corrections have been obtained in the framework of $f(R)$ gravity as a general feature of these theories [5, 6, 7, 8]. The physical meaning of such corrections needs to be confirmed at small and large scale lengths. Different experimental, geophysical and astronomical constraints on Yukawa violations of the gravitational inverse square law are given in Figs. 9 and 10 from [9] for different ranges. From these results we can conclude that the Yukawa term is relatively well constrained for the short ranges. Also, for longer distances Yukawa corrections have been successfully applied to clusters of galaxies [10, 11, 7]. Further tests are needed in order to set good constraints on Yukawa corrections, especially in the ranges of Galactic stellar dynamics.

S-stars are young stars that closely orbit the massive compact object at the center of Milky Way, named Sgr A* [12, 13, 14, 15]. These stars, together with recently discovered dense gas cloud falling towards the Galactic Centre [15], indicate that the massive central object is a probably black hole. We find astrometric observations for one of them, called S2. These observation indicate that its orbit maybe do not yield closed ellipses [12, 16].

2. Theory

A general gravitational potential, with a Yukawa correction, can be obtained in the Newtonian limit of any analytic $f(R)$-gravity model.

$$\mathcal{S} = \int d^4x\sqrt{-g}\left[f(R) + \mathcal{X}\mathcal{L}_m\right] . \tag{1}$$

The simplest extension of General Relativity is achieved assuming:

$$R \rightarrow f(R) , \qquad \omega(\phi) = 0 \tag{2}$$

One can assume analytic Taylor expandable $f(R)$ functions with respect to the value $R = 0$ that is the Minkowskian background [17]:

$$f(R) = \sum_{n=0}^{\infty} \frac{f^{(n)}(0)}{n!} R^n = f_0 + f_1 R + \frac{f_2}{2} R^2 + ... \tag{3}$$

At the order $\mathcal{O}(0)$, the field equations give the condition $f_0 = 0$ and then the solutions at further orders do not depend on this parameter. Considering the first term in R, f_0 has the meaning of a cosmological constant.

In $f(R)$-gravity, the scalar curvature R of the Hilbert - Einstein action, is replaced by a generic function $f(R)$. As a result, in the weak field limit [18], the gravitational potential is found to be Yukawa-like [19, 20]:

$$\Phi(r) = -\frac{GM}{(1+\delta)r}\left[1 + \delta e^{-\left(\frac{r}{\Lambda}\right)}\right] , \tag{4}$$

where $\Lambda^2 = -f_1/f_2$ is an arbitrary parameter (usually referred to as the range of interaction), depending on the typical scale of the system under consideration and $\delta = f_1 - 1$ is a universal constant.

In order to constrain the parameters, we simulate orbits of S2 star in Yukawa-like gravity potentials and fit them to the astrometric observations obtained by New Technology Telescope/Very Large Telescope (NTT/VLT)[12].

The simulated orbits of S2 star were obtained by numerical integration of differential equations of motion in Yukawa gravitational potential:

$$\dot{\mathbf{r}} = \mathbf{v}, \quad \mu\ddot{\mathbf{r}} = -\nabla\Phi\left(\mathbf{r}\right), \tag{5}$$

where μ is so called reduced mass in the two-body problem. In our calculations we assumed that the mass of central black hole is $M_{BH} = 4.3 \times 10^6 M_\odot$ and that the distance to the S2 star is $d_\star = 8.3$ kpc [12]. Detailed description of calculation and fitting procedure is given in the papers [21, 22, 23]. Perturbations from other members of the stellar cluster, as well as from some possibly existing extended structures composed from visible or dark matter [24, 25] were neglected due to simplicity reasons.

3. Results and discussions

Our goal is that the Yukawa-like correction, coming from $f(R)$ gravity, can be used in order to fix the coefficients in the expansion (3). For the expansion up to the second order, we have 2 parameters to fix. According to [2], the relations between f_1, f_2 and δ and Λ parameters are $f_1 = 1 + \delta$, $f_2 = -(1+\delta)/(\Lambda^2)$. We have to find the minimal values of the reduced χ^2 in order to determine f_1 and f_2 assuming $f_0 = 0$. This allows to reconstruct $f(R)$ models up to the second order.

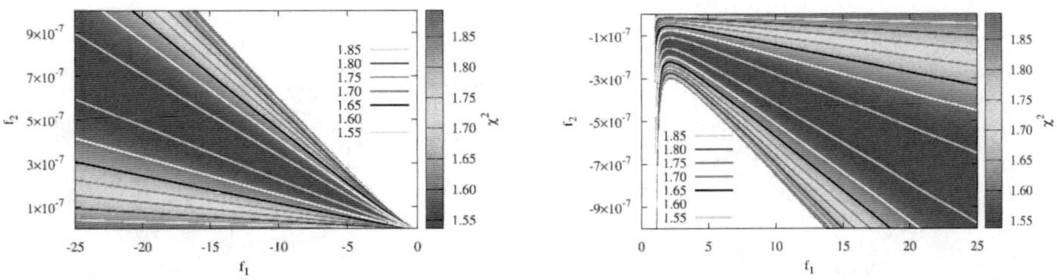

Figure 1: The maps of reduced χ^2 over the $\{f_1 - f_2\}$ parameter space of $f(R)$ gravity in case of NTT/VLT observations of S2 star which give at least the same ($\chi^2 = 1.89$) or better fits ($\chi^2 < 1.89$) than the Keplerian orbits. The left panel corresponds to f_1 in $[-25, 0]$, and the right panel to f_1 in $[0, 25]$. A few contours are presented for specific values of reduced χ^2 given in figure's legend.

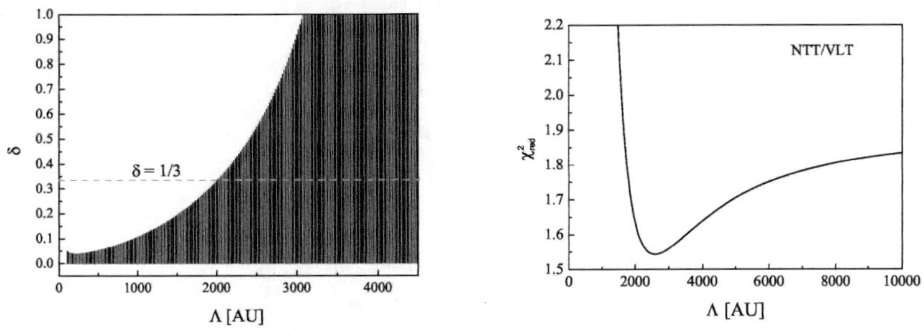

Figure 2: (Left panel) The parameter space for Yukawa gravity under the constraint that, during one orbital period, S2-like star orbits in Yukawa gravity differ less than ϵ from the corresponding orbits in Newtonian gravity for the value of parameter ϵ= 0.01. (Right panel) The reduced χ^2 for δ=1/3 as a function of Λ in case of NTT/VLT observations.

Fig. 1 presents the maps of the reduced χ^2 over the $\{f_1 - f_2\}$ parameter space for all simulated orbits of S2 star which give at least the same or better fits to the NTT/VLT observations of S2 star than the Keplerian orbits ($\chi^2 = 1.89$). In a large region of the parameter space, χ^2 of the orbits in modified potential is less than the value in Newtonian potential. It comes that we cannot constrain both f_1 and f_2 using only the observed S2 orbits because these two parameters are strongly correlated. We can con-

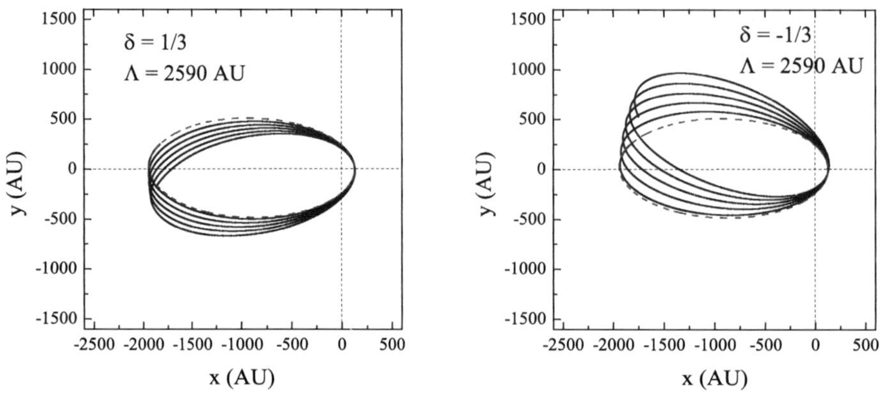

Figure 3: Comparisons between the orbit of S2 star in Newtonian gravity (red dashed line) and Yukawa gravity (blue solid line) during 5 orbital periods for $\Lambda = 2.59 \times 10^3$ AU. In the left panel $\delta = +1/3$, and in the right $\delta = -1/3$, respectively.

strain only their ratio f_1/f_2.

In Fig. 2 (left panel) we presented the parameter space for Yukawa gravity under the constraint that, during one orbital period, S2-like star orbits in Yukawa gravity differ less than ϵ from the corresponding orbits in Newtonian gravity, for the value of parameter ϵ=0.01. The value of parameter ϵ has very strong influence of constraints on the parameters of Yukawa gravity theories. Additionally, in Fig. 2 we designated value of $\delta = 1/3$ by line.

Fig. 2 (right panel) presents the reduced χ^2 for all fits with fixed value of δ=1/3 as a function of Λ which was varied from 10 to 10 000 AU. In the case of NTT/VLT observations the minimum of reduced χ^2 is 1.54 and is obtained for $\Lambda = 2.59 \times 10^3$ AU, while in the case of NTT/VLT+Keck combined data set the minimal value of 3.24 is obtained for $\Lambda = 3.03 \times 10^3$ AU. For both cases the reduced χ^2 for Keplerian orbits ($\delta = 0$) are 1.89 and 3.53, respectively. These values are significantly higher than the corresponding minima for $\delta = 1/3$. We can conclude that Yukawa gravity describes observed data even better than

Newtonian gravity and that $\delta = 1/3$ is valid value at galactic scales.

The simulated orbits of S2 star around the central object in Yukawa gravity (blue solid line) and in Newtonian gravity (red dashed line) for $\Lambda = 2.59 \times 10^3$ AU and $\delta = +1/3$ (left panel) and $\delta = -1/3$ (right panel) during 5 periods, are presented in Fig. 3. We can notice that for $\delta = -1/3$ the precession has the negative direction and when $\delta = +1/3$ the precession has the positive direction (like in General Relativity). Our analysis shows that the Yukawa gravity potential induces precession of S2 star orbit in the same direction as General Relativity for $\delta > 0$ and for $\delta < -1$, and in the opposite direction for $-1 < \delta < 0$ as in the case of extended mass distribution or in R^n gravity [21].

4. Conclusions

In this paper we constrained the parameters of Yukawa gravity using the observed positions of S2 orbit around the Galactic Centre. Our investigation shows that:

1. for vanishing δ, we recover the Keplerian orbit of S2 star;
2. the most probable value for Yukawa gravity parameter Λ in the case of S2 star, is around 5000 - 7000 AU;
3. the two parameters of Yukawa gravity Λ and δ are highly correlated in the range ($0 < \delta < 1$). For $\delta > 2$ they are not correlated;
4. if we take $\delta = +1/3$, i.e. the same universal constant δ which was successfully applied to clusters of galaxies [10, 11] and rotation curves of spiral galaxies [7], we also get a good agreement (better than in the Keplerian case) in the case of observations of S2 star orbit;

5. the scale parameter of Yukawa gravity in the case of S2 star for $\delta = +1/3$ is about: $\Lambda \approx 3000 \pm 1500$ AU;
6. for $\delta = +1/3$ there is orbital precession in positive direction like in General Relativity, and for $\delta = -1/3$ the precession has negative direction, as in the case of extended mass distribution or in R^n gravity [21];
7. the connection between coefficients f_1 and f_2 and Yukawa gravity parameters Λ and δ are following: $\Lambda^2 = -f_1/f_2$ $\delta = f_1 - 1$. We cannot constrain both f_1 and f_2 using only the observed S2 orbits because these two parameters are strongly correlated. We can constrain only their ratio f_1/f_2.

The newest astrometric data for the star S2 of NTT/VLT measurements and Keck measurements indicate that maybe the S2 orbits deviate from the Keplerian case, due to the relativistic precession. The accuracy of the observations is constantly improving from around 10 mas during the first part of the observational period, currently reaching less than 1 mas. With that currently limit one can not say for sure that S2 star orbit really deviates from the Newtonian case. We hope that future observations with advanced facilities, such as GRAVITY[26] or METIS[27], will be able to verify these claims.

References

[1] S. Capozziello and M. De Laurentis, Physics Reports **509** (2011) 167.
[2] S. Capozziello and M. De Laurentis, Ann. Phys. **524** No. 9-10 (2012) 545.
[3] S. Nojiri and S. D. Odintsov, Physics Reports **505** (2011) 59.
[4] H. A. Buchdahl, Mon. Not. R. Astron. Soc. **150** (1970) 1.
[5] K. S. Stelle, Gen. Rel. Grav. **9** (1978) 353.
[6] S. Capozziello, A. Stabile, and A. Troisi, Mod. Phys. Lett. A **24** (2009) 659.
[7] V. F. Cardone and S. Capozziello, Mon. Not. R. Astron. Soc. **414** (2011) 1301.
[8] L. Iorio, Mon. Not. R. Astron. Soc. **401** (2010) 2012.
[9] E. G. Adelberger, J. H. Gundlach, B. R. Heckel, S. Hoedl, and S. Schlamminger, Progress in Particle and Nuclear Physics **62**, (2009) 102.

[10] S. Capozziello, A. Stabile, and A. Troisi, Phys. Rev. D **76** (2007) 104019.

[11] S. Capozziello, E. de Filippis, and V. Salzano, Mon. Not. R. Astron. Soc. **394** (2009) 947.

[12] S. Gillessen, F. Eisenhauer, T. K. Fritz, H. Bartko, K. Dodds-Eden, O. Pfuhl, T. Ott, and R. Genzel, Astrophys. J. **707** (2009) L114.

[13] S. Gillessen, F. Eisenhauer, S. Trippe, T. Alexander, R. Genzel, F. Martins, and T. Ott, Astrophys. J. **692** (2009) 1075.

[14] A. M. Ghez, S. Salim, N. N. Weinberg, J. R. Lu, T. Do, J. K. Dunn, K. Matthews, M. R. Morris, S. Yelda, E. E. Becklin, T. Kremenek, M. Milosavljević, and J. Naiman, Astrophys. J. **689** (2008) 1044.

[15] S. Gillessen, R. Genzel, T. K. Fritz et al., Nature **481** (2012) 51.

[16] L. Meyer, A. M. Ghez, R. Schödel, S. Yelda, A. Boehle, J. R. Lu, T. Do, M. R. Morris, E. E. Becklin, K. Matthews, Science **338**(2012) 84.

[17] A. Stabile and S. Capozziello, Phys. Rev. D **87** (2013) 064002.

[18] T. Clifton and J. D. Barrow, Phys. Rev. D **72** (2005) 103005.

[19] S. Capozziello, V. Faraoni, *Beyond Einstein gravity: A Survey of gravitational theories for cosmology and astrophysics*, Fundamental Theories of Physics, Vol. 170, Springer, New York (2010).

[20] N. R. Napolitano, S. Capozziello, A. J. Romanowsky, M. Capaccioli, and C. Tortora, Astrophys. J. **748** (2012) 87.

[21] D. Borka, P. Jovanović, V. Borka Jovanović, and A. F. Zakharov, Phys. Rev. D **85** (2012) 124004.

[22] D. Borka, P. Jovanović, V. Borka Jovanović, and A. F. Zakharov, JCAP **11** (2013) 050.

[23] S. Capozziello, D. Borka, P. Jovanović, and V. Borka Jovanović, Phys. Rev. D **90** (2014) 044052.

[24] A. F. Zakharov, A. A. Nucita, F. De Paolis, and G. Ingrosso, Phys. Rev. D **76** (2007) 062001.

[25] A. F. Zakharov, D. Borka, V. Borka Jovanović, and P. Jovanović, J. Adv. Space Res. 54 (2014) 1108.

[26] S. Gillessen, F. Eisenhauer, G. Perrin, et al. *Proc. of SPIE*, (2010) 77342-1-9.

[27] A. Eckart, N. Sabha, G. Witzel, et al. *Proc. of SPIE* (2012) arXiv:1208.1129.

Branes at toric conical singularities*

Dmitri Bykov†
Max-Planck-Institut für Gravitationsphysik, Albert-Einstein-Institut,
Am Mühlenberg 1, D-14476 Potsdam-Golm, Germany
Steklov Mathematical Institute of Russ. Acad. Sci.,
Gubkina str. 8, 119991 Moscow, Russia

Abstract

After a brief survey of results related to the application of the AdS/CFT correspondence to $\mathcal{N} = 1$ supersymmetric models, I elaborate on certain geometric problems arising in this setup, more particularly on the construction of a Ricci-flat metric on the cone over a del Pezzo surface of rank one.

1. The AdS/CFT correspondence

The AdS/CFT conjecture states that certain conformal field theories (CFT) are in a precise way dual to string theory models describing string propagation in anti-de-Sitter space (AdS). The original example is the duality between the maximally supersymmetric Yang-Mills theory ($\mathcal{N} = 4$ SYM) and the type IIB superstring in the space $AdS_5 \times S^5$ [1]. Consider Yang-Mills theory with gauge group $SU(N)$ and coupling constant $g_{\rm YM}$. From the point of view of the $AdS_5 \times S^5$ geometry these two parameters – N and $g_{\rm YM}$ – are related to the radius of the sphere and the flux of the self-dual 5-form through it. [1]:

$$R^2 \sim \sqrt{g^2_{\rm YM} N}, \qquad \int_{S^5} F_5 \sim N. \tag{1}$$

The correspondence between the two theories manifests itself, in particular, in the fact that the global symmetries of field theory correspond to the super-isometries of $AdS_5 \times S^5$. Indeed, the superconformal group of the $\mathcal{N} = 4$ theory is $PSU(2, 2|4)$. Its maximal bosonic subgroup is $SU(2, 2) \times$

* Work supported in part by grants RFBR 14-01-00695-a, 13-01-12405 ofi-m2 and the grant MK-2510.2014.1 of the President of Russia Grant Council. I would like to thank B. Dragovich and I. Salom for the invitation to the 8-th Mathematical Physics Meeting and for their kind hospitality in Belgrade.

† e-mail address: dmitri.bykov@aei.mpg.de, dbykov@mi.ras.ru

[1] We note that, since the form F_5 is self-dual, it has nonzero components along the AdS space as well.

$SU(4)$. Here the first factor is isomorphic to $SO(2,4)$ – the conformal group of four-dimensional Minkowski space, and the second one represents the R-symmetry group of the theory. On the other hand, $SO(2,4)$ and $SO(6)$ are isometry groups of the anti-de-Sitter space and the five-sphere, respectively.

The simplest modification of this basic model is obtained if one factorizes the sphere by a discrete subgroup of the isometry group $SO(6)$:

$$AdS_5 \times S^5 \longrightarrow AdS_5 \times S^5/\Gamma, \qquad \Gamma \subset SO(6). \tag{2}$$

The dual gauge theory is now different: the gauge group consists of several simple factors, and the matter multiplets are in bifundamental representations. The number of supersymmetries is also reduced, and it depends on the structure of the group Γ.

2. Sasakian manifolds and $\mathcal{N} = 1$ supersymmetry

All manifolds of the type $AdS_5 \times S^5/\Gamma$ are particular cases of a rather wide class of solutions of IIB supergravity of the form $AdS_5 \times X^5$, in which the dilaton ϕ is constant and the only nonzero form is the 5-form:

$$\phi = \text{const.}, \qquad F_5 \sim N\left((\text{vol})_{AdS_5} + (\text{vol})_{X^5}\right). \tag{3}$$

The Bogomolnyi-Prasad-Sommerfeld condition for this configuration, i.e. the requirement of preservation of at least one supersymmetry, reduces to the vanishing of the gravitino variation:

$$\delta\psi_\mu = (\nabla_\mu + F_5\gamma_\mu)\epsilon = 0. \tag{4}$$

The spinor ϵ, satisfying (4), is called a Killing spinor. The existence of a Killing spinor substantially constrains the geometry of the space X^5. In particular, one can show, similarly to the way it is done in [2], that the solubility of the equation (4) is equivalent to the requirement that the metric cone over X^5 should be Kähler and Ricci-flat. In other words, let $(\widetilde{ds^2})_{X^5}$ be the metric on X^5. The metric on the cone is, by definition, $ds^2 = dr^2 + r^2(\widetilde{ds^2})_{X^5}$. The requirement of supersymmetry is that the metric ds^2 is Kähler and Ricci-flat (the latter following as well directly from the fact, that due to the supergravity equations of motion and the 5-form (3) $(\widetilde{ds^2})_{X^5}$ is an Einstein metric of positive curvature). In this case the metric on the base of the cone $(\widetilde{ds^2})_{X^5}$ is called Sasaki-Einstein.

Locally all Sasaki-Einstein metrics look rather simple: they can be written in the form

$$(ds^2)_{X^5} = (d\varphi - A)^2 + \sum_{i,j=1}^{2} g_{i\bar{j}}\, dz_i\, d\bar{z}_j. \tag{5}$$

Here $g_{i\bar{j}}$ is a Kähler-Einstein metric on some complex surface (the meaning of this surface is explained below, see section 2.1.1.) and A is the Kähler current. Validity of such representation is clear for the sphere S^5, if one recalls that it is the total space of a Hopf fiber bundle with base $\mathbb{CP}^2$. An important difference of the general case is that $g_{i\bar{j}}$ does not necessarily have to be smooth (although the metric on X^5 is smooth).

On a Sasaki-Einstein manifold with the metric (5) there is a canonical vector field $\partial/\partial\varphi$ with fixed norm – it is the so-called Reeb vector. From the point of view of the dual gauge theory, this $U(1)$ isometry of the space X^5 is dual to the global $U(1)$ symmetry of the superconformal field theory – the R-symmetry, which acts on the supercharges [3]:

$$Q \to e^{i\alpha}Q, \qquad \bar{Q} \to e^{-i\alpha}\bar{Q}. \tag{6}$$

In contrast to a generic theory with $\mathcal{N} = 1$ supersymmetry, a superconformal theory always possesses a $U(1)$ R-symmetry, since the generator of R-symmetry enters explicitly in the superalgebra $SU(2,2\,|\,1)$ (hence the R-transformations are its inner automorphisms).

2.1. Calabi-Yau manifolds and singularities

The construction of the cone over a Sasaki-Einstein manifold X^5, described above, allows one to take a new look at the configuration of the type $AdS_5 \times X^5$. Here one can recall the interpretation of the $\mathcal{N} = 4$ SYM theory as an effective field theory describing the oscillations of a stack of parallel D3-branes embedded in flat space $\mathbb{R}^{1,9}$. Clearly, in this case the transverse space to the branes is $\mathbb{R}^6$. The branes are massive objects, and therefore they change the geometry of the space, in which they are embedded. Assuming that the branes are located at the origin of the space $\mathbb{R}^6$, one can introduce a radial coordinate r and look for supergravity solutions of the type

$$ds^2 = h^{-1/2}(r)\sum_{i=1}^{4} dx_i^2 + h^{1/2}(r)(dr^2 + r^2(d\Omega)_{S^5}). \tag{7}$$

Since the branes are charged with respect to the 5-form F_5, there is an extra condition $\int_{S^5} F_5 \sim N$. The solution has the form [4] (l is a linear scale)

$$h(r) = 1 + \frac{l^4\,N}{r^4}. \tag{8}$$

In the limit $r \to 0$ we obtain the metric of $AdS_5 \times S^5$. It is clear then that the sphere S^5 emerges as the locus of points, equidistant from $r = 0$ (in the natural flat metric) in the six-dimensional space, in which the branes are embedded. Now we can consider the case when the 'internal' six-dimensional space is not a flat one, but rather a general (compact) Calabi-Yau space Y^6. Placing the branes at a nonsingular point of Y^6, we will again have $\mathcal{N} = 4$ SYM as a low-energy limit. The situation changes, however, if we place the branes at a singular point of the Calabi-Yau space. In this case the effective field theory on the branes depends on the local geometry of Y^6 in the vicinity of the singularity. The neighborhood of the singularity can be described by a noncompact Ricci-flat metric of the conical type, i.e. the metric which has the form $(ds^2)_{sing} = dr^2 + r^2 (\widetilde{ds^2})_{X^5}$. In this case the points equidistant from the singularity form not an S^5 but rather the Sasaki-Einstein manifold X^5. The ansatz (7) is still valid and leads in the limit $r \to 0$ to the configuration $AdS_5 \times X^5$ [5].

2.1.1. The cones

It follows from the previous section that the spaces X^5 are tightly connected with the singularities of the Calabi-Yau manifolds Y^6 of complex dimension 3. Here we will focus on those singularities which are singularities of complex cones over complex surfaces. The surfaces will be denoted by $\mathcal{M}$ hereafter. However, it is simplest to explain the notion with the example of a cone over a complex curve, namely when $\mathcal{M} = \mathbb{CP}^1$. Indeed, consider a conic in $\mathbb{CP}^2$:

$$X_0^2 + X_1^2 + X_2^2 = 0 \tag{9}$$

This algebraic variety is isomorphic to $\mathbb{CP}^1$, and it is nonsingular as a hypersurface in $\mathbb{CP}^2$. In order to pass to the cone, we should 'forget' that X_0, X_1, X_2 are projective coordinates, that is to say we will treat them as ordinary affine variables. In other words, we will consider the equation (9) as defined in $\mathbb{C}^3$. Thus one obtains a singularity at the origin $X_0 = X_1 = X_2 = 0$. We will call this singular variety the complex cone over $\mathbb{CP}^1$. This definition can be extended in a straightforward way to complex cones over complex surfaces (one just needs to consider a higher-dimensional ambient space $\mathbb{CP}^N$ and certain algebraic equations therein).

The only remaining question is in which case the affine manifolds of the sort (9) are Ricci-flat (or, more precisely, in which case they remain Ricci-flat after the singularity at the origin has been resolved). It turns out that this requirement is met if and only if the complex surface $\mathcal{M}$ has an ample anticanonical bundle. (In the example (9) one is in fact dealing with a complex curve $\mathbb{CP}^1$, which has an ample anticanonical bundle as well.) In the language of differential geometry it means that $\mathcal{M}$ is a positively curved surface, or, more exactly, that the integral of the Ricci form over any homologically nontrivial two-cycle C is positive:

$$\int_C \frac{i}{2\pi} R_{m\bar{n}}\, dz_m \wedge d\bar{z}_n > 0. \tag{10}$$

It is known from algebraic geometry that the only compact nonsingular surfaces of positive curvature are $\mathbb{CP}^1 \times \mathbb{CP}^1$, $\mathbb{CP}^2$, as well as the blow-ups of $\mathbb{CP}^2$ in no more than eight points – these are the so-called del Pezzo surfaces (two-dimensional Fano varieties).

Pick one of these surfaces $\mathcal{M}$. How can one build a corresponding Sasaki-Einstein metric on X^5? When $\mathcal{M}$ admits a Kähler-Einstein metric (with Kähler potential K), the answer is given by formula (5). It can be rewritten in a slightly different way, if one recalls the correspondence between the Kähler-Einstein metrics on $\mathcal{M}$ and Ricci-flat metrics on Y^6. One can search for a Kähler potential $\mathcal{K}$, which defines the (Ricci-flat) metric on the complex cone over $\mathcal{M}$, in the following form:

$$\mathcal{K} = \mathcal{K}(|u|^2 e^K). \tag{11}$$

This ansatz is known as Calabi's ansatz [6].

3. Cone over the del Pezzo surface

The del Pezzo surface of rank one, $\mathbf{dP}_1$, is the blow-up of $\mathbb{CP}^2$ at one point. There is a concrete algebraic model for it. The surface can be embedded into $\mathbb{CP}^8$, and the embedding is given by those sections of $\mathcal{O}_{\mathbb{CP}^2}(3)$ which vanish at a given point on $\mathbb{CP}^2$, for example at $(z_1 : z_2 : z_3) = (0 : 0 : 1)$. The embedding is called *anticanonical*, since the standard tautological sheaf $\mathcal{O}_{\mathbb{CP}^8}(1)$ over the ambient space $\mathbb{CP}^8$, when restricted to the surface, coincides with its anticanonical sheaf. Once the embedding is specified, the *affine cone* may be constructed simply by passing from projective space $\mathbb{CP}^8$ to the affine space $\mathbb{C}^9$.

In this construction we have chosen a reference point $(0 : 0 : 1) \in \mathbb{CP}^2$, which was subsequently blown-up. This reduces the automorphism group $PGL(3, \mathbb{C})$ of $\mathbb{CP}^2$ to the automorphism group of $\mathbf{dP}_1$:

$$Aut(\mathbf{dP}_1) = P\begin{pmatrix} \bullet & \bullet & 0 \\ \bullet & \bullet & 0 \\ \bullet & \bullet & \bullet \end{pmatrix}, \tag{12}$$

and the cone $Y := \mathrm{Cone}(\mathbf{dP}_1)$ has as its automorphism group the maximal parabolic subgroup of $GL(3, \mathbb{C})$ defined by matrices of the form (12) (forgetting the projectivization).

We will be looking for a Kähler metric on Y with the isometry group being the maximal compact subgroup of $Aut(Y)$:

$$\mathrm{Isom}(Y) = U(2) \times U(1) \tag{13}$$

In more practical terms, we will introduce three complex coordinates z_1, z_2, u on Y and, due to the $U(2) \times U(1)$ isometry, we will assume that the Kähler potential depends on the two combinations of them:

$$K = K(|z_1|^2 + |z_2|^2, |u|^2) \tag{14}$$

The corresponding Kähler form is $\Omega = i\partial\bar{\partial}K$ and the metric is $g_{i\bar{j}} = \partial_i \bar{\partial}_j K$. The Ricci tensor is related to the metric of a Kähler manifold as $R_{i\bar{j}} = -\partial_i \bar{\partial}_j \log \det g$. The Ricci-flatness (Calabi-Yau) condition $R_{i\bar{j}} = 0$ takes the form of a Monge-Ampere equation for the function $G(\mu, \nu)$, which is the

Legendre transform of K with respect to the variables $t = \log(|z_1|^2 + |z_2|^2)$ and $s = \log(|u|^2)$:

$$G = \mu\, t + \nu\, s - K \tag{15}$$

The usefulness of the new variables (μ, ν) to a large extent relies on the fact that they have a transparent geometric meaning – these are the moment maps for the following two $U(1)$ actions on Y:

$$U(1)_\mu : \quad (\, z_1 \to e^{i\alpha}\, z_1, \quad z_2 \to e^{i\alpha}\, z_2 \,) \qquad\qquad U(1)_\nu : \quad u \to e^{i\beta}\, u \tag{16}$$

The Ricci-flatness equation has the following form:

$$e^{\frac{\partial G}{\partial \mu} + \frac{\partial G}{\partial \nu}} \left(\frac{\partial^2 G}{\partial \mu^2} \frac{\partial^2 G}{\partial \nu^2} - \left(\frac{\partial^2 G}{\partial \mu \partial \nu} \right)^2 \right) = \tilde{a}\, \mu \tag{17}$$

Denoting (μ, ν) by (μ_1, μ_2), we can recover the metric from the dual potential G [7] using the formula

$$ds^2 = \mu\, g_{\mathbb{CP}^1} + \sum_{i,j=1}^{2} \frac{\partial^2 G}{\partial \mu_i \partial \mu_j}\, d\mu_i\, d\mu_j + \sum_{i,j=1}^{2} \left(\frac{\partial^2 G}{\partial \mu^2} \right)^{-1}_{ij} (d\phi_i - A_i)\, (d\phi_j - A_j)\,, \tag{18}$$

where $g_{\mathbb{CP}^1}$ is the standard round metric on $\mathbb{CP}^1$, $A_2 = 0$ and A_1 is the 'Kähler current' of $\mathbb{CP}^1$, i.e. a connection, whose curvature is the Fubini-Study form of $\mathbb{CP}^1$: $dA_1 = \omega_{\mathbb{CP}^1}$.

Since (μ, ν) are moment maps for the $U(1)^2$ action, the domain on which the potential $G(\mu, \nu)$ is defined is the moment polygon for this $U(1)^2$ action. In this case it is an unbounded domain with two vertices, hence we may call it a 'biangle'.

From the perspective of the equation (17), it is the singularities of the function G that determine the polygon. As we approach an arbitrary edge L_i of the polygon, i.e. when $L_i \to 0$, we impose the asymptotic condition

$$G = L_i\, (\log L_i - 1) + \ldots \quad \text{as} \quad L_i \to 0, \tag{19}$$

where the ellipsis indicates terms regular at $L_i \to 0$. Despite being subleading, they are important for the equation (17) to be consistent even in the limit $L_i \to 0$.

Notice that, in addition to the $U(1)^2$ action (16), there is yet another $U(1)$ action given by $(z_1 \to e^{i\alpha}\, z_1,\ z_2 \to e^{-i\alpha}\, z_2)$. Therefore the fiber over a generic point of the biangle is $\mathbb{CP}^1 \times \mathbb{T}^2$. The angles of the moment polygon are detemined by the normal bundles to the two $\mathbb{CP}^1$'s 'located' in the corners (see [8] for a detailed discussion).

3.1. An expansion away from the vertex of the cone

To start the analysis of the equation (17) first of all we shift the origin along the μ-axis by a constant μ_0 in such a way that the new origin is located at the intersection point of the two outer lines of the moment 'biangle'.

We aim at building an expansion of the metric at 'infinity', i.e. far from the 'vertex'. For this purpose, instead of the $\{\mu, \nu\}$ variables, we will use a 'radial' variable ν and an 'angular' variable ξ:

$$\{\mu, \nu\} \to \left\{\nu,\ \xi = \frac{\mu - \mu_0}{\nu}\right\} \tag{20}$$

Then the equation (17) above may be rewritten as follows:

$$e^{\frac{\partial G}{\partial \nu} - \frac{\xi - 1}{\nu}\frac{\partial G}{\partial \xi}} \left[\frac{\partial^2 G}{\partial \xi^2}\frac{\partial^2 G}{\partial \nu^2} - \left(\frac{\partial^2 G}{\partial \xi \partial \nu} - \frac{1}{\nu}\frac{\partial G}{\partial \xi}\right)^2\right] = a\, \nu^3 \left(\xi + \frac{\mu_0}{\nu}\right) \tag{21}$$

We propose the following expansion for the potential G at $\nu \to \infty$ (b is a constant):

$$G = 3\nu(\log \nu - 1) + \nu\, P_0(\xi) + b \log \nu + \sum_{k=0}^{\infty} \nu^{-k}\, P_{k+1}(\xi) \tag{22}$$

Substituting this expansion in the equation, we obtain a 'master' equation, which can then be expanded in powers of $\frac{1}{\nu}$ and solved iteratively for the functions $P_k(\xi)$.

3.1.1. Leading order

The first equation is obtained from (21) in the limit $\nu \to \infty$:

$$P_0'' = \frac{a}{3}\, \xi\, e^{(\xi - 1)P_0' - P_0} \tag{23}$$

and has the solution

$$P_0(\xi) = \log\left(-\frac{a}{9}\right) - \sum_{i=0}^{2} \frac{\xi - \xi_i}{\xi_i - 1} \log(\xi - \xi_i), \tag{24}$$

where ξ_i are the roots of the polynomial

$$Q(\xi) = \xi^3 - \frac{3}{2}\xi^2 + d, \tag{25}$$

and d is a constant of integration, which plays a crucial geometric role.

The function $P_0(\xi)$ determines the metric at infinity by means of the formulas (22) and (18). One can check that in the original (μ, ν) variables the 'radial' part of the metric is determined by

$$G_0 = \sum_{i=0}^{2} \frac{\mu - \xi_i\,\nu}{1-\xi_i}\,(\log(\mu - \xi_i\,\nu) - 1) \tag{26}$$

It turns out that there is a topological restriction on the roots ξ_i, which is related to the fact that the two $\mathbb{CP}^1$'s embedded in the corners of the moment 'biangle' should have correct normal bundles (a full discussion of this can be found in [8]). As a result the value of d in (25) is uniquely fixed to

$$d = \frac{16+\sqrt{13}}{64}\,. \tag{27}$$

Quite generally, it may be shown that the function G satisfying eq. (21) has the following structure (up to a certain 'regularity requirement' at the edges of the moment biangle, see [8]):

$$G = \frac{\tilde{\mu}-\xi_1\,\nu}{1-\xi_1}\left(\log\left(\frac{\tilde{\mu}-\xi_1\,\nu}{1-\xi_1}\right)-1\right) + \frac{\tilde{\mu}-\xi_2\,\nu}{1-\xi_2}\left(\log\left(\frac{\tilde{\mu}-\xi_2\,\nu}{1-\xi_2}\right)-1\right) + \\ + \left(\frac{\tilde{\mu}-\xi_0\,\nu}{1-\xi_0}+b\right)\left(\log\left(\frac{\tilde{\mu}-\xi_0\,\nu}{1-\xi_0}+b\right)-1\right) + b\sum_{k=2}^{\infty}\left(\frac{b}{\nu}\right)^k \tilde{P}_{k-2}(\xi),$$

where $\tilde{P}_k(\xi)$ is a polynomial of degree k. Here the variable μ has been shifted in such a way that the new origin is located at $\tilde{\mu} = \nu = 0$.

3.1.2. Singular points of the Heun equation and eigenfunctions

We proceed to describe in more detail the equations that arise in higher orders of perturbation theory. In the k-th order we arrive at the following equation:

$$D_k\tilde{P}_k := \frac{d}{d\xi}\left(Q(\xi)\frac{d\tilde{P}_k}{d\xi}\right) - \left((k+1)^2-1\right)\xi\tilde{P}_k = \text{r.h.s.}, \tag{28}$$

where $Q(\xi)$ has been defined in (25) and the right hand side depends on the previous orders of perturbation theory, i.e. on $\tilde{P}_{k-1}, \ldots, \tilde{P}_0$ and their derivatives. One can show that the inhomogeneous equation (28) has a polynomial solution of degree k. The general solution, however, is produced by adding to this particular solution a general solution of the homogenized equation $D_k\Pi_k = 0$. The roots $\xi_i, i = 0, 1, 2$ of the polynomial $Q(\xi) = \prod_{i=0}^{2}(\xi - \xi_i)$ are singular points of this equation. Moreover, by making the change of variables $\xi \to \frac{1}{\xi}$, one easily sees that ∞ is a singular point as well. Hence $D_k\Pi_k = 0$ is a Fuchsian equation with 4 singular points – a particular case of the so-called Heun equation, in which all exponents are zero.

The question we wish to pose is whether the homogenized equation $D_k\Pi_k = 0$ has a nontrivial solution regular at *two* of the singular points, say ξ_1, ξ_2. This is necessary in order to comply with the regularity requirement mentioned above. We claim that the answer is positive only for $k = 0, 1$:

$$\Pi_0 = \alpha \tag{29}$$

$$\Pi_1 = \beta(\xi - 1), \tag{30}$$

$$\text{where} \quad \alpha, \beta = \text{const.}$$

Quite interestingly, the nontrivial solutions are independent of the constant d, which suggests that they are also relevant for the deformations of other Ricci-flat cones asymptotic to (real) cones over Sasaki-Einstein manifolds.

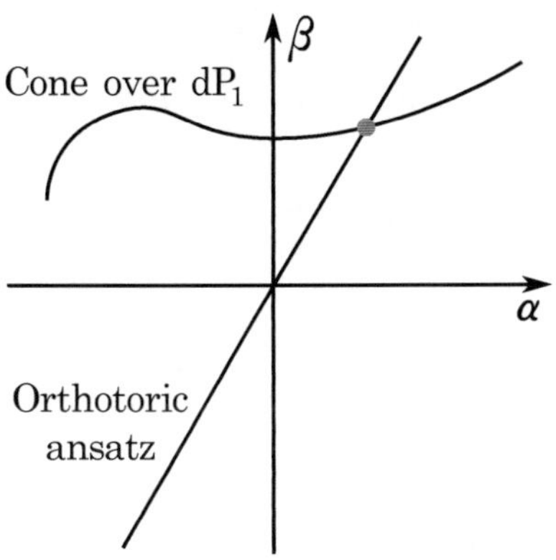

Figure 1: The red spot represents the known (orthotoric) metric on the cone over $\mathbf{dP}_1$.

In general the fact that some parameters are absent in the metric at infinity and then appear in different orders of the expansion around this metric is compatible with the known cases. One prominent example is the resolved conifold, which is asymptotic to the real cone over the Einstein-Sasaki manifold $T^{1,1} = \frac{SU(2)\times SU(2)}{U(1)}$ at infinity and exhibits two resolution parameters in an expansion around infinity [9].

As we have just discussed, there exists a Ricci-flat metric with $U(2)\times U(1)$ isometry on the complex cone over $\mathbf{dP}_1$ with *at most* two parameters, which we termed α and β. There exists a closed expression for G, and hence for the metric, in a particular case when the parameters α and β are related in a certain way — this is the metric obtained in [10], as well as in [11] by means of the so-called 'orthotoric' ansatz developed in [12].

4. Conclusion and outlook

We have explained, how remarkable geometric objects appear in the analysis of $\mathcal{N} = 1$ superconformal theories via the AdS/CFT correspondence. As a particular interesting example, we have analyzed the parameter space of Ricci-flat metrics on the complex cone over a del Pezzo surface of rank one (sometimes also called the Hirzebruch surface F_1). In particular, using an expansion at infinity, we have found one potential new parameter β (see (30)). In general we conjecture that there is a particular relation between β and α that preserves the correct topology, i.e. $\beta = \beta(\alpha)$ (see Fig. 1).

In this case the remaining parameter is related to the size of the blown-up $\mathbb{CP}^1$ in the base of the cone, i.e. in the del Pezzo surface.

It would be very interesting to obtain an exact formula (like the one of [12]-[11]), for the solution with two generic values of the parameters α, β, and this would certainly shed light on these questions.

References

[1] J. M. Maldacena, "The Large N limit of superconformal field theories and supergravity," Int. J. Theor. Phys. **38** (1999) p. 1113 [Adv. Theor. Math. Phys. **2** (1998) p. 231]

[2] C. Bär, "Real Killing spinors and holonomy," Commun. Math. Phys. **154**: 3 (1993) p.509

[3] D. Berenstein, C. P. Herzog and I. R. Klebanov, "Baryon spectra and AdS /CFT correspondence," JHEP **0206** (2002) 047

[4] G. T. Horowitz and A. Strominger, "Black strings and P-branes," Nucl. Phys. B **360** (1991) p. 197.

[5] D. R. Morrison and M. R. Plesser, "Nonspherical horizons. 1.," Adv. Theor. Math. Phys. **3** (1999) p. 1

[6] E. Calabi, "Métriques kählériennes et fibrés holomorphes," Ann. Sci. École Norm. Sup. (4) 12: 2 (1979) p. 269

[7] H. Pedersen and Y. Poon, "Hamiltonian constructions of Kähler-Einstein metrics and Kähler metrics of constant scalar curvature.," Commun. Math. Phys. **136**: 2 (1991) p. 309.

[8] D. Bykov, "Comments on the del Pezzo cone," arXiv:1405.2319

[9] L. A. Pando Zayas and A. A. Tseytlin, "3-branes on spaces with $R \times S^2 \times S^3$ topology," Phys. Rev. D **63** (2001) 086006.

[10] W. Chen, H. Lu, and C. Pope, "Kerr-de Sitter black holes with NUT charges," Nucl.Phys. B **762** (2007) p. 38.

[11] D. Martelli and J. Sparks, "Resolutions of non-regular Ricci-flat Kahler cones," J.Geom.Phys. **59** (2009) p. 1175.

[12] V. A. D. M. Calderbank and P. Gauduchon, "Hamiltonian 2-forms in Kähler geometry. I: General theory.," J. Differ. Geom. **73**: 3 (2006) p. 359.

Fundamental issues in extended geometry

Martin Cederwall*

Fundamental Physics, Chalmers Univ. of Technology, Gothenburg, Sweden

Abstract

This talk gives the basics of extended geometry, with a focus on the gauge transformations, the generalised diffeomorphisms. Some global issues are discussed. I also go into some detail about how the formalism can be made to allow for a geometric derivation of discrete duality transformations and of monodromies for non-geometric field configurations.

Duality symmetries in string theory/M-theory mix gravitational and non-gravitational fields. Manifestation of such symmetries calls for a generalisation of the concept of geometry.

It has been proposed that the compactifying space (torus) is enlarged to accommodate momenta (representing momenta and brane charges) in modules of a duality group. This leads to *doubled geometry* [1]-[23]. in the context of T-duality, and *exceptional geometry* [24]-[39] in the context of U-duality.

In the present talk, I will

- Describe the basics of extended geometry: fields, gauge transformations, &c.
- Discuss some global issues concerning generalised manifolds.
- Make precise how duality transformations become "geometric", and what remains for a full description.
- Point out some questions and directions.

* e-mail address: martin.cederwall@chalmers.se

Consider a compactification from 11 to $11 - n$ dimensions on T^n. As is well known, all fields and charges fall into modules of $E_{n(n)}$:

n	$E_{n(n)}$	R_1
3	$SL(3) \times SL(2)$	$(\mathbf{3}, \mathbf{2})$
4	$SL(5)$	**10**
5	$Spin(5,5)$	**16**
6	$E_{6(6)}$	**27**
7	$E_{7(7)}$	**56**
8	$E_{8(8)}$	**248**

I will focus on diffeomorphisms, and how they generalise. The ordinary diffeomorphisms go together with gauge transformations for the 3-form and (dual) 6-form fields (and for high enough n also gauge transformations for dual gravity) in an $E_{n(n)}$ module R_1. This is the "coordinate module". The derivative transforms in $\bar{R}_1$.

The situation for T-duality is simpler. Compactification from 10 to $10-d$ dimensions give the (continuous) T-duality group $O(d,d)$. The momenta are complemented with string windings to form the $2d$-dimensional module.

Note that the duality group is not to be seen as a global symmetry. Instead, discrete duality transformations in $O(d,d;\mathbb{Z})$ or $E_{n(n)}(\mathbb{Z})$ should arise as symmetries in certain backgrounds, just as the mapping class group $SL(n;\mathbb{Z})$ arises as discrete isometries of a torus. The rôle of the continuous versions of the duality groups should be analogous to that of $GL(n)$ in ordinary geometry (gravity).

One has to decide how tensors transform. The generic recipe is to mimic the Lie derivative for ordinary diffeomorphisms:

$$L_U V^m = U^n \partial_n V^m - \partial_n U^m V^n$$

where the first term is a transport term and the second a **gl** transformation. In the case of U-duality, the role of GL is assumed by $E_{n(n)} \times \mathbb{R}$, and

$$\begin{aligned} \mathscr{L}_U V^M &= L_U V^M + Y^{MN}{}_{PQ} \partial_N U^P V^Q \\ &= U^N \partial_N V^M + Z^{MN}{}_{PQ} \partial_N U^P V^Q \end{aligned}$$

where $Z^{MN}{}_{PQ} = -\alpha_n P^M_{\text{adj}}{}_{Q,}{}^N{}_P + \beta_n \delta^M_Q \delta^N_P = Y^{MN}{}_{PQ} - \delta^M_P \delta^N_Q$. projects on the adjoint of $E_{n(n)} \times \mathbb{R}$. Y is an invariant tensor, the form of which we do not give here (see ref. [34]).

The transformations form an algebra for $n \leq 7$:

$$[\mathscr{L}_U, \mathscr{L}_V]W^M = \mathscr{L}_{[\![U,V]\!]}W^M$$

where the "Courant bracket" is $[\![U,V]\!]^M = \frac{1}{2}(\mathscr{L}_U V^M - \mathscr{L}_V U^M)$, provided that the derivatives fulfill a "*section condition*".

This section condition ensures that fields locally depend only on an n-dimensional subspace of the coordinates, on which a $GL(n)$ subgroup acts. It reads $Y^{MN}{}_{PQ}\partial_M \ldots \partial_N = 0$, or

$$(\partial \otimes \partial)|_{\bar{R}_2} = 0$$

n	R_1	R_2
3	$(\mathbf{3},\mathbf{2})$	$(\bar{\mathbf{3}},\mathbf{1})$
4	**10**	$\bar{\mathbf{5}}$
5	**16**	**10**
6	**27**	$\overline{\mathbf{27}}$
7	**56**	**133**
8	**248**	$\mathbf{1} \oplus \mathbf{3875}$

The interpretation of the section condition is that the momenta locally are chosen so that they may span a linear subspace of cotangent space with maximal dimension, such that any pair of covectors p, p' in the subspace fulfill $(p \otimes p')|_{\bar{R}_2} = 0$.

The corresponding statement in T-duality is $\eta^{MN}\partial_M \otimes \partial_N = 0$, where η is the $O(d,d)$-invariant metric. The maximal linear subspace is a d-dimensional isotropic (light-like) subspace, and it is determined by a pure spinor Λ. Once a Λ is chosen, the section condition can be written $\Gamma^M \Lambda \partial_M = 0$. An analogous linear construction can be performed in the exceptional setting [34].

The generalised diffeomorphisms do not satisfy a Jacobi identity. On general grounds, it can be shown that the "Jacobiator" is proportional to $(\!([\![U,V]\!],W)\!) + \text{cycl}$, where $(\!(U,V)\!) = \frac{1}{2}(\mathscr{L}_U V + \mathscr{L}_V U)$.

It is important to show that the Jacobiator in some sense is trivial. It turns out that $\mathscr{L}_{(\!(U,V)\!)}W = 0$ (for $n \leq 7$), and the interpretation is that it is a gauge transformation with a parameter representing reducibility.

In doubled geometry, this reducibility is just the scalar reducibility of a gauge transformation: $\delta B_2 = d\lambda_1$, with the reducibility $\delta\lambda_1 = d\lambda'_0$.

In exceptional geometry, the reducibility turns out to be more complicated, leading to an infinite (but well defined) reducibility, containing the modules of tensor hierarchies, and providing a natural generalisation of forms (having connection-free covariant derivatives).

I will skip the detailed description of the generalised gravity. It effectively provides the local dynamics of gravity and 3-form, which are encoded by a vielbein $E_M{}^A$ in the coset $(E_{n(n)} \times \mathbb{R})/K(E_{n(n)})$.

n	$E_{n(n)}$	$K(E_{n(n)})$
3	$SL(3) \times SL(2)$	$SO(3) \times SO(2)$
4	$SL(5)$	$SO(5)$
5	$Spin(5,5)$	$(Spin(5) \times Spin(5))/\mathbb{Z}_2$
6	$E_{6(6)}$	$USp(8)/\mathbb{Z}_2$
7	$E_{7(7)}$	$SU(8)/\mathbb{Z}_2$

The T-duality case is described by a generalised metric or vielbein in $O(d,d)/(O(d) \times O(d))$, parametrised by the ordinary metric and B-field.

With some differences from ordinary geometry, one can go through the construction of connection, torsion, metric compatibility &c., and arrive at generalised Einstein's equations encoding the equations of motion for all fields.

One may introduce global or local supersymmetry, although the concept of superfields and supergeometry is quite unexplored.

Just like a manifold may be described by an atlas of coordinate charts with transition functions on the overlaps, we want to patch a generalised manifold by overlaps, that must be *finite generalised diffeomorphisms*.

In ordinary geometry, the transition functions are matrices $M_M{}^N = \frac{\partial X^N}{\partial X'^M}$, and covectors obey

$$A'_M(X') = M_M{}^N A_N(X)\,.$$

Now we need M to be replaced by a group element F in $E_{n(n)} \times \mathbb{R}$ or $O(d,d)$:

$$A'_M(X') = F_M{}^N(M) A_N(X)\,.$$

The matrix F is known explicitly for $O(d,d)$ [16][18].

$$F(M) = \frac{1}{2}\left(M(M^{-1})^t + (M^{-1})^t M\right)$$

Only partial results exist for exceptional groups [40].

The $O(d,d)$ result can be obtained from exponentiation of the generalised Lie derivative. The naïve composition rule does not hold, $F(M)F(N) \neq F(MN)$, *i.e.*, the map $F : GL(2d) \to O(d,d)$ is not a group homomorphism. Instead, a "twisted" version holds [18],

$$F(M)F(N) = F(MN)e^{\Delta},$$

where e^{Δ} is a generalised diffeomorphism that leaves the coordinated unchanged. The existence of such transformations is due to the section condition, and has no counterpart in ordinary geometry. Such a "non-translating" generalised coordinate transformation only transforms the B-field, and not the metric (given an explicit solution to the section condition).

The situation can be summarised as follows: For any choice of $M = \frac{\partial X}{\partial X'}$ there is an equivalence class of generalised diffeomorphisms, all given by $F(M,\Delta) = F(M)e^{\Delta}$ for some Δ, with $F(M)$ as a canonical representative. The map from $GL(2n)$ to the equivalence class *is* a homomorphism.

This leads to a gerbe structure. Defining

$$H(M,N) = F(M)F(M^{-1}N)F(N^{-1}),$$

the product

$$\Lambda(M,N,P) = H(M,N)H(N,P)H(P,M)$$

defines the non-trivial triple overlap (cocycle).

It is not surprising that a gerbe structure arises, given that tensor gauge transformations are contained in the formalism. It is however striking that the structure can be examined very concretely, and that the abelian gerbe is embedded in the (non-abelian) $O(d,d)$.

We expect "slightly" non-abelian gerbes to arise in the U-duality context, as soon as the 6-form dual to the 3-form becomes important, *i.e.*, for $n \geq 5$. These (finite) transformations have yet to be constructed.

Back to the geometric origin of duality symmetries. To what extent can they be obtained as "generalised isometries"? There is a severe restriction, that is a result of the section condition. The situation is analogous in the $O(d,d)$ and $E_{n(n)}$ cases, I review the $O(d,d)$ situation for simplicity.

The solution to the section condition, forcing all fields to depend only on a subset of the coordinates, identified as "ordinary" space, is not changed by generalised diffeomorphisms. Transformations in $O(d,d)$ preserving an isotropic subspace do not fill out the entire $O(d,d)$, but only $GL(d) \ltimes \wedge^2\mathbf{d}$.

In a basis with $X = (x, \tilde{x})$, where x are the physical coordinates, they take the form

$$\begin{bmatrix} m & \bullet \\ 0 & m^{-1} \end{bmatrix}$$

In a basis with $X = (x, \tilde{x})$, where x are the physical coordinates, they take the form

$$\begin{bmatrix} m & \bullet \\ 0 & m^{-1} \end{bmatrix}$$

This means that a large part of $O(d,d,\mathbb{Z})$ is excluded. Even the simple T-duality transformation interchanging momenta and string windings (or x and $\tilde{x}$) through "$R \leftrightarrow \frac{1}{R}$" cannot be obtained as a generalised diffeomorphism.

It also means that such transformations are not available as transition functions on overlaps, so that genuinely "non-geometric" solutions can not be constructed.

How can the situation be saved? It turns out that double diffeomorphisms can be formulated not only using the algebraic invariant $O(d,d)$ metric η_{MN}, but any pseudo-Riemannian (split-signature) metric $H_{MN}(X)$:

$$\mathscr{L}_\xi V_M = (L_\xi V)_M - H_{MP}H^{NQ}D_Q\xi^P V_N\,.$$

where now the covariant derivative D contains the torsion-free affine connection for H.

The potential curvature obstructions in the algebra then "miraculously" cancel [19, 23]. Some further flatness restrictions are imposed by the consistency of the covariant section condition. The defining metric H is not dynamical.

Corresponding statements should be true for exceptional groups, but there the structure in question is not a metric structure.

The point of such a "pre-geometric" formulation is that it becomes clear that any isometry of H will be a global symmetry of the model. For infinitesimal (continuous) isometries, the statement $[L_u, \mathscr{L}_\xi] = \mathscr{L}_{[u,\xi]}$ can be verified explicitly, and the analogous statement is true for finite isometries, *e.g.* the discrete isometries of a torus, where $O(d,d;\mathbb{Z})$ is the isometry subgroup of the mapping class group of the $2d$-torus.

Isometries are automorphisms of the generalised diffeomorphisms. Their parameters are not restricted by the section condition, and may therefore We may, and should, declare them as part of the gauge symmetry. By doing this, the problem of geometrising duality is solved, and a prescription is obtained for how fields transform under it. The same is true for non-geometric field configurations, where monodromy around some loop in a base space may take values in the gauge group, now containing T-duality transformations.

To conclude:

- Extended geometry unifies metric and tensor fields, and their respective gauge symmetries, in a framework providing interesting generalisations of ordinary geometry.
- Many questions remain to be examined, especially concerning finite transformations and the construction of "generalised manifolds". Most pressing is the issue of the section condition. Although it is covariant, its solutions explicitly break the continuous duality groups. It should preferably be promoted to a dynamically generated constraint, which would allow for true actions.
- A superspace formulation seems realistic for T-duality, but more problematic for U-duality. Simultaneous manifestation of supersymmetry and duality through some generalisation of pure spinor superfields may prove very powerful.
- Partial results exist on exceptional diffeomorphisms for $n > 7$. This is where dual gravity becomes important. This deserves further investigation.

References

[1] M.J. Duff, *"Duality rotations in string theory"*, Nucl. Phys. **B335** (1990) 610.

[2] A.A. Tseytlin, *"Duality symmetric closed string theory and interacting chiral scalars"*, Nucl. Phys. **B350** (1991) 395.

[3] W. Siegel, *"Two vierbein formalism for string inspired axionic gravity"*, Phys. Rev. **D47** (1993) 5453 [arXiv:hep-th/9302036].

[4] W. Siegel, *"Superspace duality in low-energy superstrings"*, Phys. Rev. **D48** (1993) 2826 [arXiv:hep-th/9305073].

[5] W. Siegel, *"Manifest duality in low-energy superstrings"*, in Berkeley 1993, Proceedings, Strings '93 353 [arXiv:hep-th/9308133].

[6] C.M. Hull, *"A geometry for non-geometric string backgrounds"*, J. High Energy Phys. **0510** (2005) 065 [arXiv:hep-th/0406102].

[7] C.M. Hull, *"Doubled geometry and T-folds"*, J. High Energy Phys. **0707** (2007) 080 [arXiv:hep-th/0605149].

[8] C. Hull and B. Zwiebach, *"Double field theory"*, J. High Energy Phys. **0909** (2009) 99 [arXiv:0904.4664].

[9] O. Hohm, C.M. Hull and B. Zwiebach, *"Background independent action for double field theory"*, J. High Energy Phys. **1007** (2010) 016 [arXiv:1003.5027].

[10] O. Hohm, C.M. Hull and B. Zwiebach, *"Generalized metric formulation of double field theory"*, J. High Energy Phys. **1008** (2010) 008 [arXiv:1006.4823].

[11] O. Hohm and S.K. Kwak, *"Frame-like geometry of double field theory"*, J. Phys. **A44** (2011) 085404 [arXiv:1011.4101].

[12] O. Hohm and S.K. Kwak, *"$N = 1$ supersymmetric double field theory"*, J. High Energy Phys. **1203** (2012) 080 [arXiv:1111.7293].

[13] I. Jeon, K. Lee and J.-H. Park, *"Differential geometry with a projection: Application to double field theory"*, J. High Energy Phys. **1104** (2011) 014 [arXiv: 1011.1324].

[14] I. Jeon, K. Lee and J.-H. Park, *"Stringy differential geometry, beyond Riemann"*, Phys. Rev. **D84** (2011) 044022 [arXiv:1105.6294].

[15] I. Jeon, K. Lee and J.-H. Park, *"Supersymmetric double field theory: stringy reformulation of supergravity"*, Phys. Rev. **D85** (2012) 081501 [arXiv:1112.0069].

[16] O. Hohm and B. Zwiebach, *"Large gauge transformations in double field theory"*, J. High Energy Phys. **1302** (2013) 075 [arXiv:1207.4198].

[17] J.-H. Park, *"Comments on double field theory and diffeomorphisms"*, J. High Energy Phys. **1306** (2013) 098 [arXiv:1304.5946].

[18] D.S. Berman, M. Cederwall and M.J. Perry, *"Global aspects of double geometry"*, J. High Energy Phys. **1409** (2014) 66 [arXiv:1401.1311].

[19] M. Cederwall, *"The geometry behind double geometry"*, J. High Energy Phys. **1409** (2014) 70 [arXiv:1402.2513].

[20] O. Hohm, D. Lüst and B. Zwiebach, *"The spacetime of double field theory: Review, remarks and outlook"*, Fortschr. Phys. **61** (2013) 926 [arXiv:1309.2977].

[21] G. Papadopoulos, *"Seeking the balance: Patching double and exceptional field theories"*, arXiv:1402.2586.

[22] C.M. Hull, *"Finite gauge transformations and geometry in double field theory"*, arXiv:1406.7794.

[23] M. Cederwall, *"T-duality and non-geometric solutions from double geometry"*, Fortschr. Phys. **12** (2014) 942 [arXiv:1409.4463].

[24] C.M. Hull, *"Generalised geometry for M-theory"*, J. High Energy Phys. **0707** (2007) 079 [arXiv:hep-th/0701203].

[25] M. Cederwall, *"M-branes on U-folds"*, in proceedings of 7th International Workshop "Supersymmetries and Quantum Symmetries" Dubna, 2007 [arXiv: 0712.4287].

[26] P.P. Pacheco and D. Waldram, *"M-theory, exceptional generalised geometry and superpotentials"*, J. High Energy Phys. **0809** (2008) 123 [arXiv:0804.1362].

[27] C. Hillmann, *"Generalized $E_{7(7)}$ coset dynamics and $D = 11$ supergravity"*, J. High Energy Phys. **0903** (2009) 135 [arXiv:0901.1581].

[28] D.S. Berman and M.J. Perry, *"Generalised geometry and M-theory"*, J. High Energy Phys. **1106** (2011) 074 [arXiv:1008.1763].

[29] D.S. Berman, H. Godazgar and M.J. Perry, *"SO(5,5) duality in M-theory and generalized geometry"*, Phys. Lett. **B700** (2011) 65 [arXiv:1103.5733].

[30] D.S. Berman, H. Godazgar, M. Godazgar and M.J. Perry, *"The local symmetries of M-theory and their formulation in generalised geometry"*, J. High Energy Phys. **1201** (2012) 012 [arXiv:1110.3930].

[31] D.S. Berman, H. Godazgar, M.J. Perry and P. West, *"Duality invariant actions and generalised geometry"*, J. High Energy Phys. **1202** (2012) 108 [arXiv: 1111.0459].

[32] A. Coimbra, C. Strickland-Constable and D. Waldram, *"$E_{d(d)} \times R^+$ generalised geometry, connections and M theory"* , J. High Energy Phys. **1402** (2014) 054 [arXiv:1112.3989].

[33] A. Coimbra, C. Strickland-Constable and D. Waldram, *"Supergravity as generalised geometry II: $E_{d(d)} \times R^+$ and M theory"*, J. High Energy Phys. **1403** (2014) 019 [arXiv:1212.1586].

[34] D.S. Berman, M. Cederwall, A. Kleinschmidt and D.C. Thompson, *"The gauge structure of generalised diffeomorphisms"*, J. High Energy Phys. **1301** (2013) 64 [arXiv:1208.5884].

[35] J.-H. Park and Y. Suh, *"U-geometry: SL(5)"*, J. High Energy Phys. **1406** (2014) 102 [arXiv:1302.1652].

[36] M. Cederwall, J. Edlund and A. Karlsson, *"Exceptional geometry and tensor fields"*, J. High Energy Phys. **1307** (2013) 028 [arXiv:1302.6736].

[37] M. Cederwall, *"Non-gravitational exceptional supermultiplets"*, J. High Energy Phys. **1307** (2013) 025 [arXiv:1302.6737].

[38] O. Hohm and H. Samtleben, *"Exceptional field theory I: $E_{6(6)}$ covariant form of M-theory and type IIB"*, Phys. Rev. **D89** (2014) 066016 [arXiv:1312.0614].

[39] O. Hohm and H. Samtleben, *"Exceptional field theory II: $E_{7(7)}$ "*, Phys. Rev. **D89** (2014) 066016 [arXiv:1312.4542].

[40] M. Cederwall and J.-H. Park, in progress.

Weakly curved background T-duals*

Lj. Davidović[†]
Institute of Physics, University of Belgrade, Serbia
B. Nikolić[‡]
Institute of Physics, University of Belgrade, Serbia
B. Sazdović[§]
Institute of Physics, University of Belgrade, Serbia

Abstract

We discuss the generalized T-dualization procedure, its connection to the standard procedure, and the results of its application to the arbitrary set of coordinates of the closed string moving in the weakly curved background. This background consists of a constant metric and linearly coordinate dependent Kalb-Ramond field with infinitesimal strength. We unite all the results into a T-dualization diagram, representing all T-dual theories, the ways to obtain the theories from one another and the T-dual coordinate transformation laws connecting the corresponding coordinates.

* Work supported in part by the Serbian Ministry of Education, Science and Technological Development, under contract No. 171031.
† e-mail address: ljubica@ipb.ac.rs
‡ e-mail address: bnikolic@ipb.ac.rs
§ e-mail address: sazdovic@ipb.ac.rs

T-duality is a symmetry seen in the string spectrum. It was observed that for toroidal compactifications [1], where one dimension is compactified on a circle of radius R and the corresponding dual dimension is compactified on a circle of radius $1/R$, one obtains the discription of a string with the same physical properties. So, there exist different string theories, describing the string in the geometrically different backgrounds, with the same predictions. Such a symmetry is not present in any point particle theory [2, 3, 4, 5], and because of that the explanation for T-duality was sought for in a fact that strings can wrap around compactified dimensions. Investigation of T-duality lead to a discovery of a Buscher T-dualization procedure [6, 7], which gave a prescription how to find the T-dual theory for same known theory. The procedure is applicable along isometry directions, what allowed the investigation of a backgrounds which do not depend on some coordinates. This procedure enabled the investigation of the properties of the background connected by T-duality. It was discovered that geometric backgrounds transform to the non-geometric backgrounds and these to different non-geometric backgrounds, which differ in a form of the background fluxes, some of which are not locally well defined [8, 9]. T-duality is also investigated for the double string theories, where T-duality is a manifest symmetry [10, 11, 12, 13].

In this talk we will discuss the results of T-dualizations done for the closed string moving in the weakly curved background, using the generalized T-dualization procedure, defined in our paper [14]. This background depends on all the space-time coordinates and as such was not a candidate for T-dualization using the standard T-dualization procedures. In paper [14], we presented generalized T-dualization procedure applicable to all space-time directions regardless of the possible background coordinate dependence. We obtained the T-dual theory which is a result of T-dualizing all the initial coordinates. In paper [15], we broadened the investigation by considering T-dualization of an arbitrary set of coordinates of both initial and its completely T-dual theory. We will recapitulate the results here and discuss further investigations. We obtained the T-dualization diagram describing the relation between all string theories T-dual to the string moving in a weakly curved background, their backgrounds and giving the T-duality laws connecting the corresponding coordinates.

So, let us start by the action describing a closed string moving in a coordinate dependent background

$$S[x] = \kappa \int_\Sigma d^2\xi \, \partial_+ x^\mu \Pi_{+\mu\nu}(x) \partial_- x^\nu, \quad \partial_\pm = \partial_\tau \pm \partial_\sigma \tag{1}$$

given in the conformal gauge $g_{\alpha\beta} = e^{2F}\eta_{\alpha\beta}$. The background field composition is defined by

$$\Pi_{\pm\mu\nu}(x) = B_{\mu\nu}(x) \pm \frac{1}{2} G_{\mu\nu}(x). \tag{2}$$

It consists of a symmetric metric tensor $G_{\mu\nu} = G_{\nu\mu}$ and an antisymmetric Kalb-Ramond field $B_{\mu\nu} = -B_{\nu\mu}$. The background must obey the following space-time equations of motion

$$R_{\mu\nu} - \frac{1}{4} B_{\mu\rho\sigma} B_{\nu}{}^{\rho\sigma} = 0, \quad D_\rho B^\rho{}_{\mu\nu} = 0, \tag{3}$$

in order to have a consistent quantum theory. We will consider one of the simplest coordinate dependent solutions, the weakly curved background, composed of a constant metric and linearly coordinate dependent Kalb-Ramond field which has an infinitesimal field strength

$$G_{\mu\nu}(x) = const,\ B_{\mu\nu}(x) = b_{\mu\nu} + \frac{1}{3} B_{\mu\nu\rho} x^\rho, \quad b_{\mu\nu}, B_{\mu\nu\rho} = const. \tag{4}$$

What are the backgrounds T-dual to this background? As, the standard T-dualization procedure is applicable to coordinate directions which do not appear as background field arguments, and the weakly curved background depends on all space-time coordinates, this procedure could not provide the answer to this question. So, a generalization of a T-dualization procedure which does not have this limitation had to be made. The main difference between the procedures obviously must be connected to background fields argument. We presented the new T-dualization procedure in [14].

Both procedures are built as a localization of a global coordinate shift symmetry $\delta x^\mu = \lambda^\mu = const$. One introduces the gauge fields v^μ_α and substitutes the ordinary derivatives with the covariant ones

$$\partial_\alpha x^\mu \to D_\alpha x^\mu = \partial_\alpha x^\mu + v^\mu_\alpha. \tag{5}$$

Imposing the following transformation law for the gauge fields

$$\delta v^\mu_\alpha = -\partial_\alpha \lambda^\mu, \quad (\lambda^\mu = \lambda^\mu(\tau, \sigma)) \tag{6}$$

one obtains that $\delta D_\alpha x^\mu = 0$. If the background does not depend on the coordinates which are T-dualized the gauge invariant action is already obtained. But, what if the background depends on all the coordinates? The additional step must be introduced. It consists of a substitution of background field argument (the coordinate x^μ), by the invariant argument (invariant coordinate) defined as a line integral of the covariant derivatives of the original coordinate

$$\Delta x^\mu_{inv} \equiv \int_P d\xi^\alpha \, D_\alpha x^\mu = x^\mu - x^\mu(\xi_0) + \Delta V^\mu, \tag{7}$$

where

$$\Delta V^\mu \equiv \int_P d\xi^\alpha v^\mu_\alpha. \tag{8}$$

Consequently the arguments of the background fields will be nonlocal. Here, they are defined as the line integrals of the gauge fields, and as such are nonlocal. Later, once the explicit form of T-dual theories is obtained the non locality will appear as dependence on double coordinates.

In order to obtain the physically equivalent theories, one must make the introduced gauge fields nonphysical which is done by requiring that there field strength

$$F^{\mu}_{\alpha\beta} \equiv \partial_\alpha v^\mu_\beta - \partial_\beta v^\mu_\alpha \tag{9}$$

must be zero. This is achieved by adding the Lagrange multiplier y_μ term to the Lagrangian. Finally, the gauge invariant action, physically equivalent to the initial action is

$$S_{inv} = \kappa \int d^2\xi \Big[D_+ x^\mu \Pi_{+\mu\nu}(\Delta x_{inv}) D_- x^\nu + \frac{1}{2}(v^\mu_+ \partial_- y_\mu - v^\mu_- \partial_+ y_\mu)\Big]. \tag{10}$$

Fixing the gauge $x^\mu(\xi) = x^\mu(\xi_0)$, one obtains

$$S_{fix}[y, v_\pm] = \kappa \int d^2\xi \Big[v^\mu_+ \Pi_{+\mu\nu}(\Delta V) v^\nu_- + \frac{1}{2}(v^\mu_+ \partial_- y_\mu - v^\mu_- \partial_+ y_\mu)\Big]. \tag{11}$$

The gauge fixed action is the main crossway of the procedure, for an appropriate equation of motion it can transform both to initial action and to the T-dual action. For the equation of motion obtained varying the action over the Lagrange multipliers $\partial_+ v^\mu_- - \partial_- v^\mu_+ = 0$, with solution $v^\mu_\pm = \partial_\pm x^\mu$, the gauge fixed action reduces to the initial action. For the equation of motion obtained varying the action over the gauge fields $\Pi_{\mp\mu\nu}[\Delta V] v^\nu_\pm + \frac{1}{2}\partial_\pm y_\mu = \mp\beta^\mp_\mu[V]$, where $\beta^\alpha_\mu[V] \equiv \partial_\mu B_{\nu\rho}\epsilon^{\alpha\beta} V^\nu \partial_\beta V^\rho$, one obtains the T-dual theory. Comparing the solutions for the gauge fields in these two directions, one obtains the T-dual coordinate transformation laws. So, for application along all directions we obtain the following connection

$$S[x] = \kappa \int_\Sigma d^2\xi\, \partial_+ x^\mu \Pi_{+\mu\nu}(x) \partial_- x^\nu \Leftrightarrow {}^\star S[y] = \frac{\kappa^2}{2} \int d^2\xi\, \partial_+ y_\mu \Theta^{\mu\nu}_-(\Delta V) \partial_- y_\nu, \tag{12}$$

with $\Delta V^\mu = V^\mu(\xi) - V^\mu(\xi_0)$, $\quad V^\mu = (g^{-1})^{\mu\nu}[(2bG^{-1})_\nu{}^\rho y_\rho + \tilde{y}_\nu]$. The dual background field composition is defined by

$$\Theta^{\mu\nu}_\pm = -\frac{2}{\kappa}(G_E^{-1}\Pi_\pm G^{-1})^{\mu\nu} = \theta^{\mu\nu} \mp \frac{1}{\kappa}(G_E^{-1})^{\mu\nu}, \tag{13}$$

and consequently the T-dual background are

$$\begin{aligned} G_{\mu\nu} &\Leftrightarrow {}^\star G^{\mu\nu}(y,\tilde{y}) = (G_E^{-1})^{\mu\nu}(\Delta V), \\ B_{\mu\nu}(x) &\Leftrightarrow {}^\star B^{\mu\nu}(y,\tilde{y}) = \frac{\kappa}{2}\theta^{\mu\nu}(\Delta V), \end{aligned} \tag{14}$$

where $G_{E\mu\nu}$ and $\theta^{\mu\nu}$ are the effective metric and the noncommutativity parameter for open bosonic string, which are

$$G_{E\mu\nu} = G_{\mu\nu} - 4(BG^{-1}B)_{\mu\nu}, \; \theta^{\mu\nu} = -\frac{2}{\kappa}(G_E^{-1}BG^{-1})^{\mu\nu}. \tag{15}$$

But what if one does not consider T-dualization over all coordinates, but only some set of coordinates. To investigate this problem, let us mark a T-dualization along direction x^μ by T^μ and a T-dualization along dual direction y_μ by T_μ. Also mark the T-dualizations along some d initial directions, all other $D-d$ initial directions, and all initial directions by

$$\mathcal{T}^a = \circ_{n=1}^{d} T^{\mu_n}, \qquad \mathcal{T}^i = \circ_{n=d+1}^{D} T^{\mu_n}, \qquad \mathcal{T} = \circ_{n=1}^{D} T^{\mu_n} \tag{16}$$

and T-dualizations along corresponding dual directions by

$$\mathcal{T}_a = \circ_{n=1}^{d} T_{\mu_n}, \qquad \mathcal{T}_i = \circ_{n=d+1}^{D} T_{\mu_n}, \qquad \tilde{\mathcal{T}} = \circ_{n=1}^{D} T_{\mu_n} \tag{17}$$

$\mu_n \in (0, 1, \dots, D-1)$. We showed in [15] that these T-dualizations form an Abelian group

$$\mathcal{T}^i \circ \mathcal{T}^a = \mathcal{T}, \quad \mathcal{T}_i \circ \mathcal{T}_a = \tilde{\mathcal{T}}, \quad \mathcal{T}_a \circ \mathcal{T}^a = 1. \tag{18}$$

We showed that all the theories T-dual to the theory of the closed bosonic string are the part of the T-dualization diagram, given by

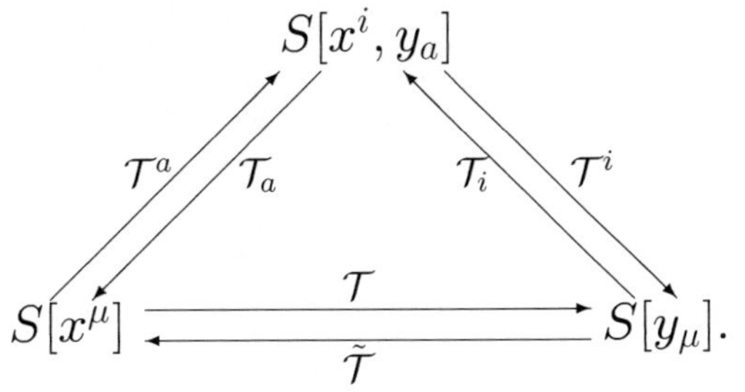

This diagram clearly describes the connection between arbitrary theory and the initial and completely T-dual theory. The explicit form of the theory obtained T-dualizing some set (marked by a) of the initial coordinates is the following

$$\begin{aligned} S[x^i, y_a] \;=\; & \kappa \int d^2\xi \Big[\partial_+ x^i \overline{\Pi}_{+ij}(x^i, \Delta V^a(x^i, y_a)) \partial_- x^j \\ & - \kappa\, \partial_+ x^i \Pi_{+ia}(x^i, \Delta V^a(x^i, y_a)) \tilde{\Theta}_-^{ab}(x^i, \Delta V^a(x^i, y_a)) \partial_- y_b \\ & + \kappa\, \partial_+ y_a \tilde{\Theta}_-^{ab}(x^i, \Delta V^a(x^i, y_a)) \Pi_{+bi}(x^i, \Delta V^a(x^i, y_a)) \partial_- x^i \\ & + \frac{\kappa}{2}\, \partial_+ y_a \tilde{\Theta}_-^{ab}(x^i, \Delta V^a(x^i, y_a)) \partial_- y_b \Big]. \end{aligned} \tag{19}$$

The new background field compositions $\overline{\Pi}_{\pm ij}$ and $\tilde{\Theta}_\pm^{ab}$ are defined as the inveres of the ordinary background field compositions $\Theta_\mp^{jk}$ and $\Pi_{\mp bc}$ reduced to the appropriate d and $D-d$ dimensional subspaces

$$\overline{\Pi}_{\pm ij} \Theta_\mp^{jk} = \Theta_\mp^{kj} \overline{\Pi}_{\pm ji} = \frac{1}{2\kappa} \delta_i^k, \tag{20}$$

$$\tilde{\Theta}^{ab}_{\pm}\Pi_{\mp bc} = \Pi_{\mp cb}\tilde{\Theta}^{ba}_{\pm} = \frac{1}{2\kappa}\delta^a_c. \tag{21}$$

It can be shown that

$$\overline{\Pi}_{+ij} \equiv \Pi_{+ij} - 2\kappa\Pi_{+ia}\tilde{\Theta}^{ab}_{-}\Pi_{+bj}. \tag{22}$$

The argument of the background fields is

$$\begin{aligned}
\Delta V^{(0)a}(x^i, y_a) &= -\kappa\Big[\tilde{\Theta}^{ab}_{0+}\Pi_{0-bi} + \tilde{\Theta}^{ab}_{0-}\Pi_{0+bi}\Big]\Delta x^{(0)i} \\
&- \kappa\Big[\tilde{\Theta}^{ab}_{0+}\Pi_{0-bi} - \tilde{\Theta}^{ab}_{0-}\Pi_{0+bi}\Big]\Delta\tilde{x}^{(0)i} \\
&- \frac{\kappa}{2}\Big[\tilde{\Theta}^{ab}_{0+} + \tilde{\Theta}^{ab}_{0-}\Big]\Delta y^{(0)}_b - \frac{\kappa}{2}\Big[\tilde{\Theta}^{ab}_{0+} - \tilde{\Theta}^{ab}_{0-}\Big]\Delta\tilde{y}^{(0)}_b,
\end{aligned} \tag{23}$$

where $\Delta x^\mu(\xi) = x^\mu(\xi) - x^\mu(\xi_0)$ and $\Delta y_\mu(\xi) = y_\mu(\xi) - y_\mu(\xi_0)$ while $\Delta\tilde{x}^\mu(\xi)$ and $\Delta\tilde{y}_\mu(\xi)$ are their duals, defined by

$$\Delta\tilde{x}^\mu(\xi) = \int_P d\xi^\alpha\, \varepsilon^\beta{}_\alpha\, \partial_\beta x^\mu, \quad \Delta\tilde{y}(\xi) = \int_P d\xi^\alpha\, \varepsilon^\beta{}_\alpha\, \partial_\beta y_\mu. \tag{24}$$

Calculating the symmetric and antisymmetric part of the background fields we obtain the T-dual metric and Kalb-Ramond field:

$$\begin{aligned}
&{}^\bullet G_{ij} = \overline{G}_{ij} = G_{ij} - G_{ia}(\tilde{G}^{-1}_E)^{ab}G_{bj} \\
&\quad -2\kappa\Big(B_{ia}\tilde{\theta}^{ab}G_{bj} + G_{ia}\tilde{\theta}^{ab}B_{bj}\Big) - 4B_{ia}(\tilde{G}^{-1}_E)^{ab}B_{bj} \\
&{}^\bullet B_{ij} = \overline{B}_{ij} = B_{ij} - \frac{\kappa}{2}G_{ia}\tilde{\theta}^{ab}G_{bj} - B_{ia}(\tilde{G}^{-1}_E)^{ab}G_{bj} \\
&\quad -G_{ia}(\tilde{G}^{-1}_E)^{ab}B_{bj} - 2\kappa B_{ia}\tilde{\theta}^{ab}B_{bj} \\
&{}^\bullet G^{ab} = (\tilde{G}^{-1}_E)^{ab} \\
&{}^\bullet B^{ab} = \frac{\kappa}{2}\tilde{\theta}^{ab} \\
&{}^\bullet G^a{}_i = \kappa\tilde{\theta}^{ab}G_{bi} + 2(\tilde{G}^{-1}_E)^{ab}B_{bi} \\
&{}^\bullet B^a{}_i = \kappa\tilde{\theta}^{ab}B_{bi} + \frac{1}{2}(\tilde{G}^{-1}_E)^{ab}G_{bi}.
\end{aligned} \tag{25}$$

As the constituents of the dual background field there appear the effective metric in the d-dimensional subspace a, defined by

$$\tilde{G}_{Eab} \equiv G_{ab} - 4B_{ac}(\tilde{G}^{-1})^{cd}B_{db}, \tag{26}$$

the non-commutativity parameter in the same subspace

$$\tilde{\theta}^{ab} \equiv -\frac{2}{\kappa}(\tilde{G}^{-1}_E)^{ac}B_{cd}(\tilde{G}^{-1})^{db}, \tag{27}$$

which combined give the new theta function $\tilde{\Theta}^{ab}_{\pm} = \tilde{\theta}^{ab} \mp \frac{1}{\kappa}(\tilde{G}_E^{-1})^{ab}$.

Every arrow in the T-duality diagram is accompanied with the appropriate T-dual coordinate transformation law. These are obtained comparing the solutions for the gauge fields in a T-dualization procedures performed between two actions in both directions. The laws for transitions

$$\mathcal{T}^a : S[x^\mu] \to S[x^i, y_a], \quad \mathcal{T}_a : S[x^i, y_a] \to S[x^\mu],$$

which are inverse to each other, are given by

$$\begin{aligned} \partial_\mp x^a &\cong -2\kappa\tilde{\Theta}^{ab}_{\mp}(x^i, \Delta V^a(x^i, y_a)) \cdot \\ &\quad \cdot \Big[\Pi_{\pm bi}(x^i, \Delta V^a(x^i, y_a))\partial_\mp x^i + \frac{1}{2}\partial_\mp y_b \mp \beta^\pm_b(x^i, V^a(x^i, y_a))\Big] \\ x^{(0)a} &\cong V^{(0)a}(x^i, y_a) \end{aligned} \tag{28}$$

and its inverse

$$\begin{aligned} \partial_\mp y_a &\cong -2\Pi_{\pm a\mu}(x)\partial_\mp x^\mu \pm 2\beta^\pm_a(x), \\ y^{(0)}_a &\cong U^{(0)}_a(x). \end{aligned} \tag{29}$$

These relations enable an investigation of the closed string non-commutativity and other geometric properties of the T-dual backgrounds. One can determine the geometric structure for an arbitrary sigma model in a T-duality diagram, find the connection between the Poisson structures of T-dual theories and the relations between non-commutativity parameters. The coordinates of the closed string are commutative when the string moves in a constant background. In a three dimensional space with the Kalb-Ramond field depending on one of the coordinates, successive T-dualizations along isometry directions lead to a theory with Q flux and the non-commutative coordinates [16, 17, 18]. Using the generalized T-dualization procedure, we found the non-commutativity characteristics of a closed string moving in the weakly curved background [15] comparing the initial and completely T-daul theory. One can expect the further investigations will reveal novelties regarding the form of the fluxes of all T-dual background forming a diagram. For now it is known that all fluxes are of type R.

References

[1] L.Brink, M.B. Green, J.H. Schwarz, *Nucl. Phys.* **B 198** (1982) 474; K. Kikkawa and M.Yamasaki, *Phys. Lett.* **B 149** (1984) 357; N. Sakai and I. Senda, *Prog. Theor. Phys.* **75** (1984) 692.

[2] A. Giveon, M. Porrati, E. Rabinovici, Phys. Rep. **244**, 77-202 (1994).

[3] E. Alvarez, L. Alvarez-Gaume and Y. Lozano, Nucl. Phys. Proc. Suppl. **41**, 1 (1995).

[4] B. Zwiebach, *A First Course in String Theory*, Second Edition, Cambridge University Press (2009).

[5] K. Becker, M. Becker and J. H. Schwarz, *String Theory and M-Theory - A Modern Introdution*, Cambridge University Press (2007).

[6] T. H. Buscher, Phys. Lett. **B 201** No. 4, 466 (1988).

[7] M. Roček and E. Verlinde, Nucl. Phys. **B 373**, 630 (1992).

[8] J. Shelton, W. Taylor and B. Wecht, JHEP **10**, 085 (2005).

[9] R. Blumenhagen, A. Deser, D. Lüst, E. Plauschinn, F. Rennecke, J. Phys. **A 44**, 385401 (2011).

[10] O. Hohm, Prog. Theor. Phys. Suppl. **188**, 116-125 (2011).

[11] C. Hull and B. Zwiebach, JHEP **09**, 099 (2009).

[12] O. Hohm, C. Hull and B. Zwiebach, JHEP **08**, 008 (2010).

[13] O. Hohm, C. Hull and B. Zwiebach, JHEP **07**, 016 (2010).

[14] Lj. Davidović and B. Sazdović, EPJ**C 74**(1), 2683 (2014).

[15] Lj. Davidović, B. Nikolić and B. Sazdović, *T-duality diagram for a weakly curved background*, arXiv:1406.5364 [hep-th].

[16] D. Lüst, JHEP **12**, 084 (2010).

[17] D. Andriot, M. Larfors, D. Lüst, P. Patalong, JHEP **06**, 021 (2013).

[18] D. Andriot, O. Hohm, M. Larfors, D. Lüst, P. Patalong, Phys. Rev. Lett. **108** , 261602 (2012).

Closed string noncommutativity in the weakly curved background*

Lj. Davidović, B. Nikolić and B. Sazdović†
Institute of Physics, University of Belgrade
11001 Belgrade, P.O.Box 57, Serbia

ABSTRACT

We consider the closed bosonic string moving in the weakly curved background. Using T-duality transformation laws we calculate the Poisson brackets of the coordinates in the T-dual space assuming that initial theory is geometric one, which means that standard Poisson algebra is obeyed. The result is that the commutative initial theory is equivalent to the non-commutative T-dual theory. All noncommutativity parameters are infinitesimal and proportional to the $B_{\mu\nu\rho}$, field strength of Kalb-Ramond field $B_{\mu\nu}$. In addition we find the algebra of the T-dual winding numbers and momenta in terms of the winding numbers and momenta of the initial theory.

1. Introduction

In order to obtain noncommutativity in the open string case it is enough to consider the open string in the presence of the *constant* gravitational $G_{\mu\nu}$ and Kalb-Ramond field $B_{\mu\nu}$ and use the boundary conditions [1, 2]. Treating boundary conditions as canonical constraints and solving them, one gets the initial coordinates expressed in terms of the Ω even effective coordinates and momenta, where Ω is world-sheet parity transformation $\Omega : \sigma \to -\sigma$. Because effective variables have nonzero Poisson bracket (PB), the PB between initial coordinates is also nonzero. The noncommutativity parameter is proportional to the Kalb-Ramond field $B_{\mu\nu}$.

There is one interesting thing which we noted in the open string case. The effective metric and the noncommutativity parameter are (up to some constants) the backgroud fields of the T-dual theory. As we know T-dual theory is physically equivalent to the initial one in the sense they have the same degrees of freedom - one at the scale R and the T-dual one at the scale $1/R$. The mathematical realization of the T-duality goes through Buscher procedure [3]. As a result of the procedure we get the relation between initial and T-dual variables which we call *transformation laws.*

* Work supported in part by the Serbian Ministry of Education and Science, under contract number No.171031.

† e-mail address: ljubica, bnikolic, sazdovic@ipb.ac.rs

The closed strings do not have endpoints, so in the constant background there are no boundary conditions. To obtain noncommutativity in the closed string case we have to use T-duality as a helping tool. But, in the constant background case, T-duality relates σ-derivatives of the coordinates of one theory with the momenta of its T-dual one. Assuming that momenta of the initial theory commute (geometric theory) it follows that the T-dual coordinates commute as well. Consequently, in the constant background case there is no closed string non-commutativity.

It is obvious that T-duality is just one part of the solution in order to get the closed string noncommutativity. The second part is coordinate dependent background obeying the space-time field equations [4, 5]. Considering the closed string in the constant gravitational field $G_{\mu\nu}$ and Kalb-Ramond field depending on one coordinate, the closed string non-commutativity was first observed in the paper [6], and investigated further in [7, 8, 9]. In these articles 3-torus is considered, where $B_{\mu\nu}$ depends on one coordinate and T-dualization is performed along two other coordinates (isometry directions) using standard Buscher procedure [3].

One can ask if it is possible to do that in the background where $B_{\mu\nu}$ depends on all space-time coordinates. The answer is affirmative but in order to achieve that we have to use the generalized T-duality procedure presented in details in [10] and to apply it to the weakly curved background. The weakly curved background used in the present article is defined by constant gravitational $G_{\mu\nu} = const$ and the linear Kalb-Ramond field $B_{\mu\nu} = b_{\mu\nu} + \frac{1}{3}B_{\mu\nu\rho}x^\rho$, where the field strength $B_{\mu\nu\rho}$ is supposed to be infinitesimal. Such background obeys space-time field equations [4, 5] in the linear approximation in $B_{\mu\nu\rho}$.

We perform the generalized T-dualization procedure [10] along all the coordinates and obtain the T-duality transformation law, $\partial_\pm y_\mu = \partial_\pm y_\mu(\partial_\pm x)$, where $\partial_\pm$ are world-sheet partial derivatives. Using canonical formalism, the T-dual coordinates are expressed in terms of the original variables, $y'_\mu \cong \frac{1}{\kappa}\pi_\mu - \beta^0_\mu[x]$, where π_μ are canonically conjugated momenta to the coordinates x^μ. The infinitesimal expression β^0_μ is the correction in comparison to the flat background case. Assuming that the coordinates and momenta of the original theory satisfy standard Poisson algebra (initial theory is geometric one), we get the coordinate noncommutativity relations in the T-dual picture. In addition, we obtain the complete algebra of the T-dual winding numbers and momenta.

2. Generalized T-duality and noncommutativity

We consider the closed bosonic string moving in the D-dimensional space-time described by the action

$$S[x] = \kappa \int_\Sigma d^2\xi\, \partial_+ x^\mu \left(B_{\mu\nu}[x] + \frac{1}{2}G_{\mu\nu}[x]\right)\partial_- x^\nu\,, \tag{1}$$

where the light-cone coordinates are defined as $\xi^\pm = \frac{1}{2}(\tau \pm \sigma)$ and the corresponding derivatives $\partial_\pm = \partial_\tau \pm \partial_\sigma$. In order to keep conformal invariance on the quantum level, the background fields have to obey the following one-loop consistency conditions [4, 5]

$$R_{\mu\nu} - \frac{1}{4}B_{\mu\rho\sigma}B_{\nu}{}^{\rho\sigma} = 0, \quad D_\rho B^\rho{}_{\mu\nu} = 0\,. \tag{2}$$

Here $B_{\mu\nu\rho} = \partial_\mu B_{\nu\rho} + \partial_\nu B_{\rho\mu} + \partial_\rho B_{\mu\nu}$ is the field strength of the field $B_{\mu\nu}$, and $R_{\mu\nu}$ and D_μ are Ricci tensor and the covariant derivative with respect to the space-time metric.

The solution of the equations in the first order in $B_{\mu\nu\rho}$, so called the weakly curved background, [7, 10, 11, 12], is defined by

$$\begin{aligned} G_{\mu\nu}[x] &= const, \\ B_{\mu\nu}[x] &= b_{\mu\nu} + h_{\mu\nu}[x] = b_{\mu\nu} + \frac{1}{3}B_{\mu\nu\rho}x^\rho, \qquad b_{\mu\nu}, B_{\mu\nu\rho} = const. \end{aligned} \tag{3}$$

Here, the field strength $B_{\mu\nu\rho}$ is infinitesimal.

Applying the generalized T-dualization procedure [10] on the closed string propagating in the weakly curved background, we obtain the T-dual action

$${}^\star S[y] = \frac{\kappa^2}{2}\int d^2\xi\; \partial_+ y_\mu \Theta_-^{\mu\nu}[\Delta V[y]]\partial_- y_\nu, \tag{4}$$

where

$$\begin{aligned} \Theta_\pm^{\mu\nu} &\equiv -\frac{2}{\kappa}(G_E^{-1}\Pi_\pm G^{-1})^{\mu\nu} = \theta^{\mu\nu} \mp \frac{1}{\kappa}(G_E^{-1})^{\mu\nu}\,, \\ G_{E\mu\nu} &\equiv G_{\mu\nu} - 4(BG^{-1}B)_{\mu\nu}\,, \quad \Pi_{\pm\mu\nu} = B_{\mu\nu} \pm \frac{1}{2}G_{\mu\nu}\,. \end{aligned} \tag{5}$$

The argument ΔV is defined nonlocally as

$$\Delta V^\mu[y] = -\kappa\theta_0^{\mu\nu}\Delta y_\nu + (g^{-1})^{\mu\nu}\Delta\tilde{y}_\nu\,, \tag{6}$$

where

$$\Delta y_\mu = \int_P (d\tau \dot{y}_\mu + d\sigma y'_\mu) = y_\mu(\xi) - y_\mu(\xi_0), \quad \Delta\tilde{y}_\mu = \int_P (d\tau y'_\mu + d\sigma \dot{y}_\mu), \tag{7}$$

and

$$g_{\mu\nu} = G_{\mu\nu} - 4(bG^{-1}b)_{\mu\nu}, \quad \theta_0^{\mu\nu} = -\frac{2}{\kappa}(g^{-1}bG^{-1})^{\mu\nu}\,. \tag{8}$$

It is obvious from the definitions (7) that these two coordinates are related by the following expressions, $\dot{y}_\mu = \tilde{y}'_\mu$, $y'_\mu = \dot{\tilde{y}}_\mu$.

The transformation laws connecting initial and T-dual coordinates play the key role in our considerations. To be more precise, we obtain from T-dualization procedure the relations between world-sheet derivatives of the initial and T-dual coordinates

$$\partial_\pm x^\mu \cong -\kappa\Theta_\pm^{\mu\nu}[\Delta V]\Big[\partial_\pm y_\nu \pm 2\beta_\nu^\mp[V]\Big], \tag{9}$$

where

$$\beta^\pm_\mu[x] = \frac{1}{2}(\beta^0_\mu \pm \beta^1_\mu) = \mp\frac{1}{2}h_{\mu\nu}[x]\partial_\mp x^\nu\,,$$
$$\beta^0_\mu[x] = h_{\mu\nu}[x]x'^\nu,\ \beta^1_\mu[x] = -h_{\mu\nu}[x]\dot{x}^\nu\,. \tag{10}$$

Because we use the canonical formalism, we must have these transformation laws in the canonical form

$$x'^\mu \cong \frac{1}{\kappa}{}^\star\pi^\mu - \kappa\theta_0^{\mu\nu}\beta^0_\nu[V], \tag{11}$$
$$\pi_\mu \cong \kappa y'_\mu + \kappa\beta^0_\mu[V]\,, \tag{12}$$

where π_μ and ${}^\star\pi^\mu$ are canonically conjugated momenta to the coordinates x^μ and y_μ, respectively. It is shown in Ref. [10] that the T-dual of the T-dual action is the initial one. If we want to have T-dual coordinates in terms of the initial ones, we just have to invert the relation (9)

$$\partial_\pm y_\mu \cong -2\Pi_{\mp\mu\nu}[\Delta x]\partial_\pm x^\nu \mp 2\beta^\mp_\mu[x]. \tag{13}$$

The canonical form of the T-dual transformations is

$$y'_\mu \cong \frac{1}{\kappa}\pi_\mu - \beta^0_\mu[x], \tag{14}$$
$${}^\star\pi^\mu \cong \kappa x'^\mu + \kappa^2\theta_0^{\mu\nu}\beta^0_\nu[x]. \tag{15}$$

Our intention is to calculate the PB's of the T-dual variables y_μ and $\tilde{y}_\mu$ using PB algebra of the initial variables. Consequently, we assume that initial theory is geometric which means that coordinates x^μ and momenta π_ν satisfy standard PB algebra

$$\{x^\mu(\sigma), \pi_\nu(\bar\sigma)\} = \delta^\mu_\nu\delta(\sigma-\bar\sigma), \quad \{x^\mu(\sigma), x^\nu(\bar\sigma)\} = 0, \quad \{\pi_\mu(\sigma), \pi_\nu(\bar\sigma)\} = 0. \tag{16}$$

In this article we will calculate, besides already mentioned PB algebra of the T-dual coordinates, also the algebra of the T-dual winding numbers and momenta. For both purposes, the first step is introducing the quantity

$$\Delta Y_\mu(\sigma,\sigma_0) = \int_{\sigma_0}^{\sigma} d\eta\, Y'_\mu(\eta) = Y_\mu(\sigma) - Y_\mu(\sigma_0), \tag{17}$$

where $Y_\mu = (y_\mu, \tilde{y}_\mu)$. The second step is to calculate their PB's. It is obvious that key relation which we have to calculate is PB between σ derivatives of Y's. When we calculate it in three possible cases it turns out that it can be written in the form

$$\{X'_\mu(\sigma), Y'_\nu(\bar\sigma)\} \cong K'_{\mu\nu}(\sigma)\delta(\sigma-\bar\sigma) + L_{\mu\nu}(\sigma)\delta'(\sigma-\bar\sigma)\,. \tag{18}$$

Integrating this relation by parts over σ and $\bar{\sigma}$, after straightforward calculation, we extract PB we are searching for

$$\{X_\mu(\tau,\sigma), Y_\nu(\tau,\bar{\sigma})\} \cong -\left[K_{\mu\nu}(\sigma) - K_{\mu\nu}(\bar{\sigma}) + L_{\mu\nu}(\bar{\sigma})\right]\theta(\sigma-\bar{\sigma})\,, \tag{19}$$

where $\theta(\sigma)$ is the step function defined as

$$\theta(\sigma) = \begin{cases} 0 & \text{if } \sigma = 0 \\ 1/2 & \text{if } 0 < \sigma < 2\pi, \\ 1 & \text{if } \sigma = 2\pi \end{cases} \qquad \sigma \in [0, 2\pi]. \tag{20}$$

This is a general form of the relation. Using transforation laws we calculate PB's in three cases: $\{y'_\mu(\sigma), y'_\nu(\bar{\sigma})\}$, $\{y'_\mu(\sigma), \tilde{y}'_\nu(\bar{\sigma})\}$ and $\{\tilde{y}'_\mu(\sigma), \tilde{y}'_\nu(\bar{\sigma})\}$, and express them in the form of (18). Reading the corresponding values of K and L and using (19), we get the noncommutativity relations for T-dual closed string coordinates

$$\{y_\mu(\sigma), y_\nu(\bar{\sigma})\} \cong -\frac{1}{\kappa}B_{\mu\nu\rho}[x^\rho(\sigma) - x^\rho(\bar{\sigma})]\theta(\sigma-\bar{\sigma}), \tag{21}$$

$$\{y_\mu(\sigma), \tilde{y}_\nu(\bar{\sigma})\} \cong -\Big\{\frac{1}{\kappa}B_{\mu\nu\rho}[\tilde{x}^\rho(\sigma) - \tilde{x}^\rho(\bar{\sigma})] - \frac{3}{2\kappa}\Gamma^E_{\rho,\mu\nu}[x^\rho(\sigma) - x^\rho(\bar{\sigma})]$$
$$+\frac{1}{\kappa}g_{\mu\nu} - \frac{3}{2\kappa}\Gamma^E_{\rho,\mu\nu}\, x^\rho(\bar{\sigma})\Big\}\theta(\sigma-\bar{\sigma}), \tag{22}$$

$$\{\tilde{y}_\mu(\sigma), \tilde{y}_\nu(\bar{\sigma})\} \cong -\Big\{-\frac{1}{\kappa}\Big[B_{\mu\nu\rho} - 6g_{\mu\alpha}Q^{\alpha\beta}{}_\rho g_{\beta\nu}\Big][x^\rho(\sigma) - x^\rho(\bar{\sigma})] \tag{23}$$
$$+\Big[-\frac{3}{2\kappa}\Big(\Gamma^E_{\mu,\nu\rho} - \Gamma^E_{\nu,\mu\rho}\Big) + \frac{4}{\kappa}B_{\mu\nu\sigma}(G^{-1}b)^\sigma{}_\rho\Big][\tilde{x}^\rho(\sigma) - \tilde{x}^\rho(\bar{\sigma})]\Big\}\theta(\sigma-\bar{\sigma})\,,$$

where

$$\tilde{x}'^\mu = \frac{1}{\kappa}(G^{-1})^{\mu\nu}\pi_\nu + 2(G^{-1}B)^\mu{}_\nu x'^\nu\,. \tag{24}$$

Here the infinitesimal fluxes are defined as

$$\Gamma^E_{\mu,\nu\rho} = \frac{1}{2}\Big(\partial_\nu G^E_{\mu\rho} + \partial_\rho G^E_{\mu\nu} - \partial_\mu G^E_{\nu\rho}\Big) = -\frac{4}{3}\Big(B_{\mu\sigma\nu}(G^{-1}b)^\sigma{}_\rho + B_{\mu\sigma\rho}(G^{-1}b)^\sigma{}_\nu\Big), \tag{25}$$

$$Q^{\mu\nu}{}_\rho = -\frac{1}{3}\Big[(g^{-1})^{\mu\sigma}(g^{-1})^{\nu\tau} - \kappa^2\theta_0^{\mu\sigma}\theta_0^{\nu\tau}\Big]B_{\sigma\tau\rho}. \tag{26}$$

For $\sigma = \bar{\sigma}$ we obtain that all PB's vanish, and consequently, coordinates commute. Also we can consider $\sigma = \bar{\sigma} + 2\pi$, which is the same point on the world-sheet as our first choice $\sigma = \bar{\sigma}$. Taking $\sigma = \bar{\sigma} + 2\pi$, three noncommutativity relations take the form

$$\{y_\mu(\sigma + 2\pi), y_\nu(\sigma)\} \cong -\frac{2\pi}{\kappa}B_{\mu\nu\rho}N^\rho\,, \tag{27}$$

$$\{y_\mu(\sigma+2\pi),\tilde{y}_\nu(\sigma)\}+\{y_\mu(\sigma),\tilde{y}_\nu(\sigma+2\pi)\}\cong -\frac{4\pi}{\kappa^2}B_{\mu\nu\rho}p^\rho$$
$$+\quad \frac{\pi}{\kappa}\left(3\Gamma^E_{\rho,\mu\nu}-8B_{\mu\nu\lambda}b^\lambda{}_\rho\right)N^\rho, \tag{28}$$

and

$$\{\tilde{y}_\mu(\sigma+2\pi),\tilde{y}_\nu(\sigma)\}\cong$$
$$\cong\quad \frac{2\pi}{\kappa}\left[-B_{\mu\nu\rho}-6g_{\mu\alpha}Q^{\alpha\beta}{}_\rho g_{\beta\nu}+2B_{\mu\nu}{}^\lambda g_{\lambda\rho}+3\left(\Gamma^E_{\mu,\nu\lambda}-\Gamma^E_{\nu,\mu\lambda}\right)b^\lambda{}_\rho\right]N^\rho$$
$$+\quad \frac{\pi}{\kappa^2}\left[3\left(\Gamma^E_{\mu,\nu\rho}-\Gamma^E_{\nu,\mu\rho}\right)p^\rho-8B_{\mu\nu\lambda}b^\lambda{}_\rho\right]p^\rho\,, \tag{29}$$

where $N^\mu=\frac{1}{2\pi}\left[x^\mu(\sigma+2\pi)-x^\mu(\sigma)\right]$ is winding number of the initial coordinates and

$$p_\mu=\frac{1}{2\pi}\int_\sigma^{\sigma+2\pi}d\eta\pi_\mu(\eta)\,, \tag{30}$$

is mean value of the momentum π_μ. Note that all three PB's are proportional to the Kalb-Ramond field strength which means they are infinitesimal.

In addition we can obtain the algebra of the T-dual winding number and momenta defined as

$$\Delta y_\mu(2\pi,0)=2\pi^\star N_\mu\,,\quad \Delta\tilde{y}_\mu(2\pi,0)=2\pi^\star P_\mu\,, \tag{31}$$

while we introduced earlier

$$\Delta x^\mu(2\pi,0)=2\pi N^\mu\,,\quad \Delta\tilde{x}^\mu(2\pi,0)=2\pi P^\mu\,. \tag{32}$$

Using (17), (18), transformation laws and above definitions we have

$$\{{}^\star N_\mu,{}^\star N_\nu\}=\frac{1}{\pi\kappa}B_{\mu\nu\rho}N^\rho\,, \tag{33}$$

$$\{{}^\star N_\mu,{}^\star P_\nu\}=\frac{1}{\pi\kappa}B_{\mu\nu\rho}P^\rho-\frac{3}{4\pi\kappa}\Gamma^E_{\rho,\mu\nu}N^\rho\,, \tag{34}$$

$$\begin{aligned}\{{}^\star P_\mu,{}^\star P_\nu\} &= -\frac{1}{\pi\kappa}\left(B_{\mu\nu\rho}-6g_{\mu\alpha}Q^{\alpha\beta}{}_\rho g_{\beta\nu}\right)N^\rho\\ &+\frac{1}{\pi}\left[-\frac{3}{2\kappa}(\Gamma^E_{\mu,\nu\rho}-\Gamma^E_{\nu,\mu\rho})+\frac{4}{\kappa}B_{\mu\nu\sigma}(G^{-1}b)^\sigma{}_\rho\right]P^\rho\,.\end{aligned}$$

3. Concluding remarks

In the present article we considered the theory describing the closed bosonic string moving in the weakly curved background and derived the non-commutativity relations using canonical approach.

We applied generalized T-duality procedure and obtained the transformation laws connecting the initial and T-dual variables. They, expressed in the canonical form, have the central role in calculation of the PB's of the T-dual coordinates y_μ and $\tilde{y}_\mu$. Infinitesimal Kalb-Ramond field strength, as a part of the function β_μ, gives the main contribution to the noncommutativity parameters. The result is that we showed the physical equivalence of the commutative initial theory and noncommutative T-dual one in linear approximation in the field strength $B_{\mu\nu\rho}$.

The general structure of the non-commutativity relations is

$$\{Y_\mu(\sigma), Y_\nu(\bar{\sigma})\} = \{F_{\mu\nu\rho}\,[x^\rho(\sigma) - x^\rho(\bar{\sigma})] + \tilde{F}_{\mu\nu\rho}\,[\tilde{x}^\rho(\sigma) - \tilde{x}^\rho(\bar{\sigma})]\}\theta(\sigma - \bar{\sigma}), \tag{35}$$

where $Y_\mu = (y_\mu, \tilde{y}_\nu)$ and $F_{\mu\nu\rho}$ and $\tilde{F}_{\mu\nu\rho}$ are the constant and infinitesimally small fluxes. At the same points, for $\sigma = \bar{\sigma}$ all PB's are zero. In the important particular case for $\sigma = \bar{\sigma} + 2\pi$ we get

$$\{Y_\mu(\sigma + 2\pi), Y_\nu(\sigma)\} = 2\pi\left[(F_{\mu\nu\rho} + 2\tilde{F}_{\mu\nu\alpha}b^\alpha_\rho)N^\rho + \frac{1}{\kappa}\tilde{F}_{\mu\nu}{}^\rho p_\rho\right], \tag{36}$$

where N^μ and p_μ are winding numbers and momenta of the original theory. In addition we calculated the PB algebra of the T-dual winding numbers and momenta in terms of the initial ones.

References

[1] F. Ardalan, H. Arfaei and M. M. Sheikh-Jabbari, *JHEP* **02** (1999) 016; C. S. Chu and P. M. Ho, *Nucl. Phys.* **B550** (1999) 151; N. Seiberg and E. Witten, *JHEP* **09** (1999) 032; F. Ardalan, H. Arfaei and M. M. Sheikh-Jabbari *Nucl. Phys.* **B576** (2000) 578; C. S. Chu and P. M. Ho, *Nucl. Phys.* **B568** (2000) 447; T.Lee, *Phys. Rev.* **D62** (2000) 024022.

[2] B. Sazdović, Eur. Phys. J. **C44** (2005) 599; B. Nikolić and B. Sazdović, *Phys. Rev.* **D 74** (2006) 045024; B. Nikolić and B. Sazdović, *Phys. Rev.* **D 75** (2007) 085011; B. Nikolić and B. Sazdović, *Adv. Theor. Math. Phys.* **14** (2010) 1, Lj. Davidovic and B. Sazdovic, *Phys. Rev.* **D 83** (2011) 066014; *JHEP* **08** (2011)112.

[3] T. H. Buscher, *Phys. Lett.* **B194** (1987) 59; T. H. Buscher, *Phys. Lett.* **B201** (1988) 466.

[4] E. S. Fradkin and A. A. Tseytlin, *Phys. Lett.* **B 158** (1985) 316; *Nucl. Phys.* **B 261** (1985) 1.

[5] K. Becker, M. Becker and J. Schwarz *String Theory and M-Theory: A Modern Introduction*, Cambridge University Press 2006; B. Zwiebach, *A First Course in String Theory*, Cambridge University Press, 2004.

[6] D. Luest, *JHEP* **12** (2010) 084.

[7] D. Andriot, M. Larfors, D. Luest, P. Patalong, *JHEP* **06** (2013) 021.

[8] D. Andriot, O. Hohm, M. Larfors, D. Luest, P. Patalong, *Phys. Rev. Lett.* 108 (2012) 261602.

[9] D. Luest, arxiv:1205.0100 [hep-th]; R. Blumenhagen, A. Deser, D. Luest, E. Plauschinn, and F. Rennecke, *J.Phys.* **A44** (2011) 385401; C. Condeescu, I. Florakis, and D. Luest, *JHEP* **04** (2012) 121.

[10] Lj. Davidović and B. Sazdović, arXiv:1205.1991 [hep-th].

[11] V. Schomerus, *Class. Quant. Grav.* **19** (2002) 5781; L.Cornalba and R.Schiappa, *Commun. Math. Phys.* **225** (2002) 33.

[12] Lj. Davidović and B. Sazdović, *Phys. Rev.* **D 83** (2011) 066014; *JHEP* **08** (2011) 112; *EPJ* **C 72** No. 11 (2012) 2199.

Positive Energy Unitary Irreducible Representations of the Superalgebras $osp(1|2n, I\!R)$ and Character Formulae

V.K. Dobrev*
Institute of Nuclear Research and Nuclear Energy,
Bulgarian Academy of Sciences,
72 Tsarigradsko Chaussee, 1784 Sofia, Bulgaria

I. Salom†
Institute of Physics, University of Belgrade,
Pregrevica 118, 11080 Zemun, Belgrade, Serbia

Abstract

We continue the study of positive energy (lowest weight) unitary irreducible representations of the superalgebras $osp(1|2n, I\!R)$. We update previous results and present the full list of these UIRs. We give also some character formulae for these representations.

1. Introduction

Recently, superconformal field theories in various dimensions are attracting more interest, in particular, due to their duality to AdS supergravities. Until recently only those for $D \leq 6$ were studied since in these cases the relevant superconformal algebras satisfy [1] the Haag-Lopuszanski-Sohnius theorem [2]. Thus, such classification was known only for the $D = 4$ superconformal algebras $su(2, 2/N)$ [3] (for $N = 1$), [4–7] (for arbitrary N). More recently, the classification for $D = 3$ (for even N), $D = 5$, and $D = 6$ (for $N = 1, 2$) was given in [8] (some results are conjectural), and then the $D = 6$ case (for arbitrary N) was finalized in [9].

On the other hand the applications in string theory require the knowledge of the UIRs of the conformal superalgebras for $D > 6$. Most prominent role play the superalgebras $osp(1|2n)$. Initially, the superalgebra $osp(1|32)$ was put forward for $D = 10$ [10]. Later it was realized that $osp(1|2n)$ would fit any dimension, though they are minimal only for $D = 3, 9, 10, 11$ (for $n = 2, 16, 16, 32$, resp.) [11]. In all cases we need to find first the UIRs of $osp(1|2n, I\!R)$ which study was started in [12] and [13].

In the present paper we intend to finalize unitarity classification of [12] and in addition to provide some character formulae.

* e-mail address: dobrev@inrne.bas.bg
† e-mail address: isalom@phy.bg.ac.rs

Since this paper is a sequel of [12], where there is extensive literature, and for the lack of space we only update the supersymmetry literature (for $D > 2$) after 2004, cf. [14–61]

2. Representations of the superalgebras $osp(1|2n)$ and $osp(1|2n, I\!R)$

2.1. The setting

Our basic references for Lie superalgebras are [62, 63], although in this exposition we follow [12].

The even subalgebra of $\mathcal{G} = osp(1|2n, I\!R)$ is the algebra $sp(2n, I\!R)$ with maximal compact subalgebra $\mathcal{K} = u(n) \cong su(n) \oplus u(1)$.

We label the relevant representations of $\mathcal{G}$ by the signature:

$$\chi = [\, d\, ;\, a_1\, , ..., a_{n-1}\,] \tag{1}$$

where d is the conformal weight, and $a_1, ..., a_{n-1}$ are non-negative integers which are Dynkin labels of the finite-dimensional UIRs of the subalgebra $su(n)$ (the simple part of $\mathcal{K}$).

In [12] were classified (with some omissions to be spelled out below) the positive energy (lowest weight) UIRs of $\mathcal{G}$ following the methods used for the $D = 4, 6$ conformal superalgebras, cf. [4–7, 9], resp. The main tool was an adaptation of the Shapovalov form [64] on the Verma modules V^χ over the complexification $\mathcal{G}^{I\!\!C} = osp(1|2n)$ of $\mathcal{G}$.

2.2. Root systems

We recall some facts about $\mathcal{G}^{I\!\!C} = osp(1|2n)$ (denoted $B(0, n)$ in [62]) as used in [12]. The root systems are given in terms of $\delta_1 \ldots, \delta_n$, $(\delta_i, \delta_j) = \delta_{ij}$, $i, j = 1, ..., n$. The even and odd roots systems are [62]:

$$\begin{aligned} \Delta_{\bar{0}} &= \{\pm\delta_i \pm \delta_j\, ,\ 1 \leq i < j \leq n\, ,\quad \pm 2\delta_i\, ,\ 1 \leq i \leq n\}\, , \\ \Delta_{\bar{1}} &= \{\pm\delta_i\, ,\ 1 \leq i \leq n\} \end{aligned} \tag{2}$$

(we remind that the signs $\pm$ are not correlated). We shall use the following distinguished simple root system [62]:

$$\Pi = \{\, \delta_1 - \delta_2\, , \ldots, \delta_{n-1} - \delta_n\, , \delta_n\, \}\, , \tag{3}$$

or introducing standard notation for the simple roots:

$$\begin{aligned} \Pi &= \{\, \alpha_1\, , ..., \alpha_n\, \}\, , \\ &\alpha_j = \delta_j - \delta_{j+1}\, , \quad j = 1, ..., n-1\, , \quad \alpha_n = \delta_n\, . \end{aligned} \tag{4}$$

The root $\alpha_n = \delta_n$ is odd, the other simple roots are even. The Dynkin diagram is:

(5)

The black dot is used to signify that the simple odd root is not nilpotent, otherwise a gray dot would be used [62]. In fact, the superalgebras $B(0,n) = osp(1|\,2n)$ have no nilpotent generators unlike all other types of basic classical Lie superalgebras [62].

The corresponding to Π positive root system is:

$$\Delta^+_{\bar{0}} = \{\delta_i \pm \delta_j\ ,\ 1 \le i < j \le n\ ,\ 2\delta_i\ ,\ 1 \le i \le n\}\ , \qquad \Delta^+_{\bar{1}} = \{\delta_i\ ,\ 1 \le i \le n\} \tag{6}$$

We record how the elementary functionals are expressed through the simple roots:

$$\delta_k = \alpha_k + \cdots + \alpha_n\ . \tag{7}$$

From the point of view of representation theory more relevant is the restricted root system, such that:

$$\bar{\Delta}^+ = \bar{\Delta}^+_{\bar{0}} \cup \Delta^+_{\bar{1}}\ , \tag{8}$$

$$\bar{\Delta}^+_{\bar{0}} \equiv \{\alpha \in \Delta^+_{\bar{0}} \mid \tfrac{1}{2}\alpha \notin \Delta^+_{\bar{1}}\} = \{\delta_i \pm \delta_j\ ,\ 1 \le i < j \le n\} \tag{9}$$

The superalgebra $\mathcal{G} = osp(1|\,2n, I\!\!R)$ is a split real form of $osp(1|\,2n)$ and has the same root system.

The above simple root system is also the simple root system of the complex simple Lie algebra B_n (dropping the distinction between even and odd roots) with Dynkin diagram:

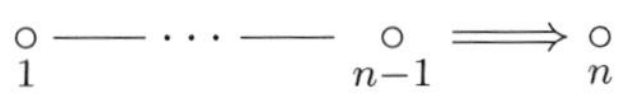

(10)

Naturally, for the B_n positive root system we drop the roots $2\delta_i$

$$\Delta^+_{\mathrm{B_n}} = \{\delta_i \pm \delta_j\ ,\ 1 \le i < j \le n\ ,\ \ \delta_i\ ,\ 1 \le i \le n\} \cong \bar{\Delta}^+ \tag{11}$$

This shall be used essentially below.

2.3. Lowest weight through the signature

Besides (1) we shall use the Dynkin-related labelling:

$$(\Lambda, \alpha_k^\vee) = -\,a_k\ ,\quad 1 \le k \le n\ , \tag{12}$$

where $\alpha_k^\vee \equiv 2\alpha_k/(\alpha_k,\alpha_k)$, and the minus signs are related to the fact that we work with lowest weight Verma modules (instead of the highest weight modules used in [63]) and to Verma module reducibility w.r.t. the roots α_k (this is explained in detail in [6, 12]).

Obviously, a_n must be related to the conformal weight d which is a matter of normalization so as to correspond to some known cases. Thus, our choice is:

$$a_n = -\,2d - a_1 - \cdots - a_{n-1}\ . \tag{13}$$

The actual Dynkin labelling is given by:

$$m_k \;=\; (\rho - \Lambda, \alpha_k^\vee) \tag{14}$$

where $\rho \in \mathcal{H}^*$ is given by the difference of the half-sums $\rho_{\bar 0}\,, \rho_{\bar 1}$ of the even, odd, resp., positive roots (cf. (6):

$$\begin{aligned}\rho \;\doteq\; \rho_{\bar 0} - \rho_{\bar 1} \;&=\; (n-\tfrac12)\delta_1 + (n-\tfrac32)\delta_2 + \cdots + \tfrac32\delta_{n-1} + \tfrac12\delta_n \ , \\ \rho_{\bar 0} \;&=\; n\delta_1 + (n-1)\delta_2 + \cdots + 2\delta_{n-1} + \delta_n \ , \\ \rho_{\bar 1} \;&=\; \tfrac12(\delta_1 + \cdots + \delta_n) \ .\end{aligned} \tag{15}$$

Naturally, the value of ρ on the simple roots is 1: $(\rho, \alpha_i^\vee) = 1,\ i = 1, ..., n$.

Unlike $a_k \in \mathbb{Z}_+$ for $k < n$ the value of a_n is arbitrary. In the cases when a_n is also a non-negative integer, and then $m_k \in \mathbb{N}$ $(\forall k)$ the corresponding irreps are the finite-dimensional irreps of $\mathcal{G}$ (and of B_n).

Having in hand the values of Λ on the basis we can recover them for any element of $\mathcal{H}^*$.

We shall need only $(\Lambda, \beta^\vee)$ for all positive roots β as given in [12]:

$$\begin{aligned}(\Lambda, (\delta_i - \delta_j)^\vee) \;&=\; (\Lambda, \delta_i - \delta_j) \;=\; -\,a_i - \cdots - a_{j-1} \\ (\Lambda, (\delta_i + \delta_j)^\vee) \;&=\; (\Lambda, \delta_i + \delta_j) \;=\; 2d \;+\; a_1 + \cdots + a_{i-1} - a_j - \cdots - a_{n-1} \\ (\Lambda, \delta_i^\vee) \;&=\; (\Lambda, 2\delta_i) \;=\; 2d \;+\; a_1 + \cdots + a_{i-1} - a_i - \cdots - a_{n-1} \\ (\Lambda, (2\delta_i)^\vee) \;&=\; (\Lambda, \delta_i) \;=\; d \;+\; \tfrac12(a_1 + \cdots + a_{i-1} - a_i - \cdots - a_{n-1})\end{aligned} \tag{16}$$

2.4. Verma modules

To introduce Verma modules we use the standard triangular decomposition:

$$\mathcal{G}^{\mathbb{C}} \;=\; \mathcal{G}^+ \oplus \mathcal{H} \oplus \mathcal{G}^- \tag{17}$$

where $\mathcal{G}^+$, $\mathcal{G}^-$, resp., are the subalgebras corresponding to the positive, negative, roots, resp., and $\mathcal{H}$ denotes the Cartan subalgebra.

We consider lowest weight Verma modules, so that $V^\Lambda \cong U(\mathcal{G}^+) \otimes v_0\,$, where $U(\mathcal{G}^+)$ is the universal enveloping algebra of $\mathcal{G}^+$, and v_0 is a lowest weight vector v_0 such that:

$$\begin{aligned}Z\, v_0 \;&=\; 0\ , \quad Z \in \mathcal{G}^- \\ H\, v_0 \;&=\; \Lambda(H)\, v_0\ , \quad H \in \mathcal{H}\ .\end{aligned} \tag{18}$$

Further, for simplicity we omit the sign $\otimes$, i.e., we write $p\,v_0 \in V^\Lambda$ with $p \in U(\mathcal{G}^+)$.

Adapting the criterion of [63] (which generalizes the BGG-criterion [65] to the super case) to lowest weight modules, one finds that a Verma module V^Λ is reducible w.r.t. the positive root β iff the following holds [12]:

$$(\rho - \Lambda, \beta^\vee) = m_\beta\ , \qquad \beta \in \Delta^+\ , \quad m_\beta \in \mathbb{N}\ . \tag{19}$$

If a condition from (19) is fulfilled then V^Λ contains a submodule which is a Verma module $V^{\Lambda'}$ with shifted weight given by the pair m, β : $\Lambda' = \Lambda + m\beta$. The embedding of $V^{\Lambda'}$ in V^Λ is provided by mapping the lowest weight vector v'_0 of $V^{\Lambda'}$ to the **singular vector** $v_s^{m,\beta}$ in V^Λ which is completely determined by the conditions:

$$\begin{aligned} X\, v_s^{m,\beta} &= 0\,, \quad X \in \mathcal{G}^-\,, \\ H\, v_s^{m,\beta} &= \Lambda'(H)\, v_0\,, \quad H \in \mathcal{H}\,, \quad \Lambda' = \Lambda + m\beta\,. \end{aligned} \tag{20}$$

Explicitly, $v_s^{m,\beta}$ is given by a polynomial in the positive root generators [6, 66]:

$$v_s^{m,\beta} = P^{m,\beta}\, v_0\,, \quad P^{m,\beta} \in U(\mathcal{G}^+)\,. \tag{21}$$

Thus, the submodule I^β of V^Λ which is isomorphic to $V^{\Lambda'}$ is given by $U(\mathcal{G}^+)\, P^{m,\beta}\, v_0$.

Note that the Casimirs of $\mathcal{G}^{\mathbb{C}}$ take the same values on V^Λ and $V^{\Lambda'}$.

Certainly, (19) may be fulfilled for several positive roots (even for all of them). Let Δ_Λ denote the set of all positive roots for which (19) is fulfilled, and let us denote: $\tilde{I}^\Lambda \equiv \cup_{\beta\in\Delta_\Lambda} I^\beta$. Clearly, $\tilde{I}^\Lambda$ is a proper submodule of V^Λ. Let us also denote $F^\Lambda \equiv V^\Lambda/\tilde{I}^\Lambda$.

Further we shall use also the following notion. The singular vector v_1 is called **descendant** of the singular vector $v_2 \notin \mathbb{C}v_1$ if there exists a homogeneous polynomial P_{12} in $U(\mathcal{G}^+)$ so that $v_1 = P_{12}\, v_2$. Clearly, in this case we have: $I^1 \subset I^2$, where I^k is the submodule generated by v_k.

The Verma module V^Λ contains a unique proper maximal submodule I^Λ ($\supseteq \tilde{I}^\Lambda$) [63, 65]. Among the lowest weight modules with lowest weight Λ there is a unique irreducible one, denoted by L_Λ, i.e., $L_\Lambda = V^\Lambda/I^\Lambda$. (If V^Λ is irreducible then $L_\Lambda = V^\Lambda$.)

It may happen that the maximal submodule I^Λ coincides with the submodule $\tilde{I}^\Lambda$ generated by all singular vectors. This is, e.g., the case for all Verma modules if rank $\mathcal{G} \leq 2$, or when (19) is fulfilled for all simple roots (and, as a consequence for all positive roots). Here we are interested in the cases when $\tilde{I}^\Lambda$ is a proper submodule of I^Λ. We need the following notion.

Definition: [65, 67, 68] *Let V^Λ be a reducible Verma module. A vector $v_{\rm ssv} \in V^\Lambda$ is called a* **subsingular vector** *if $v_{\rm su} \notin \tilde{I}^\Lambda$ and the following holds:*

$$X\, v_{\rm su} \in \tilde{I}^\Lambda\,, \quad \forall X \in \mathcal{G}^- \tag{22}$$

Going from the above more general definitions to $\mathcal{G}$ we recall that in [12] it was established that from (19) follows that the Verma module

$V^{\Lambda(\chi)}$ is reducible if one of the following relations holds (following the order of (16):

$$\begin{aligned} I\!N \ni m^-_{ij} &= j-i+a_i+\cdots+a_{j-1} && \text{(23a)}\\ I\!N \ni m^+_{ij} &= 2n-i-j+1+a_j+\cdots+a_{n-1}-a_1-\cdots-a_{i-1}-2d && \text{(23b)}\\ I\!N \ni m_i &= 2n-2i+1+a_i+\cdots+a_{n-1}-a_1+\cdots-a_{i-1}-2d && \text{(23c)}\\ I\!N \ni m_{ii} &= n-i+\tfrac{1}{2}(1+a_i+\cdots+a_{n-1}-a_1+\cdots-a_{i-1})-d\,. && \text{(23d)} \end{aligned}$$

Further we shall use the fact from [12] that we may eliminate the reducibilities and embeddings related to the roots $2\delta_i$. Indeed, since $m_i = 2m_{ii}$, whenever (23d) is fulfilled also (23c) is fulfilled.

For further use we introduce notation for the root vector $X^+_j \in \mathcal{G}^+$, $j=1,\ldots,n$, corresponding to the simple root α_j. Naturally, $X^-_j \in \mathcal{G}^-$ corresponds to $-\alpha_j$.

Further, we notice that all reducibility conditions in (23a) are fulfilled. In particular, for the simple roots from those condition (23a) is fulfilled with $\beta \to \alpha_i = \delta_i - \delta_{i+1}$, $i=1,...,n-1$ and $m^-_i \equiv m^-_{i,i+1} = 1+a_i$. The corresponding submodules $I^{\alpha_i} = U(\mathcal{G}^+)\, v^i_s$, where $\Lambda_i = \Lambda + m^-_i\alpha_i$ and $v^i_s = (X^+_i)^{1+a_i}\, v_0$. These submodules generate an invariant submodule which we denote by $I^\Lambda_c \subset \tilde{I}^\Lambda$. Since these submodules are nontrivial for all our signatures in the question of unitarity instead of V^Λ we shall consider also the factor-modules:

$$F^\Lambda_c = V^\Lambda / I^\Lambda_c \supset F^\Lambda\,. \tag{24}$$

We shall denote the lowest weight vector of F^Λ_c by $|\Lambda_c\rangle$ and the singular vectors above become null conditions in F^Λ_c :

$$(X^+_i)^{1+a_i}\,|\Lambda_c\rangle = 0\,, \quad i=1,...,n-1. \tag{25}$$

If the Verma module V^Λ is not reducible w.r.t. the other roots, i.e., (23b,c,d) are not fulfilled, then $F^\Lambda_c = F^\Lambda$ is irreducible and is isomorphic to the irrep L_Λ with this weight.

In fact, for the factor-modules reducibility is controlled by the value of d, or in more detail:

The maximal d coming from the different possibilities in (23b) are obtained for $m^+_{ij} = 1$ and they are:

$$d_{ij} \equiv n + \tfrac{1}{2}(a_j+\cdots+a_{n-1}-a_1-\cdots-a_{i-1}-i-j)\,, \tag{26}$$

the corresponding root being $\delta_i + \delta_j$.

The maximal d coming from the different possibilities in (23c,d), resp., are obtained for $m_i = 1$, $m_{ii} = 1$, resp., and they are:

$$\begin{aligned} d_i &\equiv n-i+\tfrac{1}{2}(a_i+\cdots+a_{n-1}-a_1-\cdots-a_{i-1})\,, && \text{(27)}\\ d_{ii} &= d_i - \tfrac{1}{2}\,, \end{aligned}$$

the corresponding roots being $\delta_i\,, 2\delta_j$, resp.

There are some orderings between these maximal reduction points [12]:

$$\begin{aligned} d_1 &> d_2 > \cdots > d_n \,, \\ d_{i,i+1} &> d_{i,i+2} > \cdots > d_{in} \,, \\ d_{1,j} &> d_{2,j} > \cdots > d_{j-1,j} \,, \\ d_i &> d_{jk} > d_\ell \,, \qquad i \le j < k \le \ell \,. \end{aligned} \tag{28}$$

Obviously the first reduction point is:

$$d_1 = n - 1 + \tfrac{1}{2}(a_1 + \cdots + a_{n-1}) \,. \tag{29}$$

3. Unitarity

The first results on the unitarity were given in [12]. These were not complete so the statement below should be called *Dobrev-Zhang-Salom Theorem.*

Theorem: All positive energy unitary irreducible representations of the superalgebras $osp(1|2n, I\!R)$ characterized by the signature χ in (1) are obtained for real d and are given as follows:

$$d \ge n - 1 + \tfrac{1}{2}(a_1 + \cdots + a_{n-1}) = d_1 \,, \quad a_1 \neq 0 \,, \tag{30}$$

$$\begin{aligned}
d &\ge n - \tfrac{3}{2} + \tfrac{1}{2}(a_2 + \cdots + a_{n-1}) = d_{12} \,, \quad a_1 = 0,\ a_2 \neq 0 \,, \\
d &= n - 2 + \tfrac{1}{2}(a_2 + \cdots + a_{n-1}) = d_2 > d_{13} \,, \quad a_1 = 0,\ a_2 \neq 0 \,, \\
d &\ge n - 2 + \tfrac{1}{2}(a_3 + \cdots + a_{n-1}) = d_2 = d_{13} \,, \quad a_1 = a_2 = 0,\ a_3 \neq 0 \,, \\
d &= n - \tfrac{5}{2} + \tfrac{1}{2}(a_3 + \cdots + a_{n-1}) = d_{23} > d_{14} \,, \quad a_1 = a_2 = 0,\ a_3 \neq 0 \,, \\
d &= n - 3 + \tfrac{1}{2}(a_3 + \cdots + a_{n-1}) = d_3 = d_{24} > d_{15} \,, \quad a_1 = a_2 = 0,\ a_3 \neq 0 \,, \\
&\cdots \\
&\cdots \\
d &\ge n - 1 - \kappa + \tfrac{1}{2}(a_{2\kappa+1} + \cdots + a_{n-1}) \,, \quad a_1 = ... = a_{2\kappa} = 0,\ a_{2\kappa+1} \neq 0 \,, \\
&\qquad \kappa = \tfrac{1}{2}, 1, ..., \tfrac{1}{2}(n-1) \,, \\
d &= n - \tfrac{3}{2} - \kappa + \tfrac{1}{2}(a_{2\kappa+1} + \cdots + a_{n-1}) \,, \quad a_1 = ... = a_{2\kappa} = 0,\ a_{2\kappa+1} \neq 0 \,, \\
&\cdots \\
d &= n - 1 - 2\kappa + \tfrac{1}{2}(a_{2\kappa+1} + \cdots + a_{n-1}) \,, \quad a_1 = ... = a_{2\kappa} = 0,\ a_{2\kappa+1} \neq 0 \,, \\
&\cdots \\
&\cdots \\
d &\ge \tfrac{1}{2}(n-1) \,, \quad a_1 = ... = a_{n-1} = 0 \\
d &= \tfrac{1}{2}(n-2) \,, \quad a_1 = ... = a_{n-1} = 0 \\
&\cdots \\
d &= \tfrac{1}{2} \,, \quad a_1 = ... = a_{n-1} = 0 \\
d &= 0 \,, \quad a_1 = ... = a_{n-1} = 0
\end{aligned} \tag{31}$$

Proof: The statement of the Theorem for $d > d_1$ is clear in [12] from the general considerations since this is the First reduction point. For $d = d_1$ (also following [12]) we have the first zero norm state which is naturally given by the corresponding singular vector $v^1_{\delta_1} = \mathcal{P}^{1,\delta_1} v_0$. In fact, all states of the embedded submodule $V^{\Lambda+\delta_1}$ built on $v^1_{\delta_1}$ have zero norms. Due to the above singular vector we have the following additional null condition in F^Λ_c :

$$\mathcal{P}^{1,\delta_1} \widetilde{|\Lambda\rangle} = 0 . \tag{32}$$

The above condition factorizes the submodule built on $v^1_{\delta_1}$. There are no other vectors with zero norm at $d = d_1$ since by a general result [63], the elementary embeddings between Verma modules are one-dimensional. Thus, F^Λ is the UIR $L_\Lambda = F^\Lambda$.
Below $d < d_1$ there is no unitarity for $a_1 \neq 0$. On the other hand (as shown in [12]) for $a_1 = 0$ the singular vector $v^1_{\delta_1}$ is descendant of the compact root singular vector $X^+_1 v_0$ which is already factored out for $a_1 = 0$. Thus, below we set $a_1 = 0$.

The next reducibility point is $d = d_{12} = n - \frac{3}{2} + \frac{1}{2}(a_2 + \cdots + a_{n-1})$. The corresponding root is $\delta_1 + \delta_2 = \alpha_1 + 2\alpha_2 + \cdots + 2\alpha_n$. The corresponding singular vector is $v^1_{\delta_1+\delta_2} = \mathcal{P}^{1,\delta_1+\delta_2} v_0$. All states of the embedded submodule $V^{\Lambda+\delta_1+\delta_2}$ built on $v^1_{\delta_1+\delta_2}$ have zero norms for $d = d_{12}$. Due to the above singular vector we have the following additional null condition in F^Λ_c :

$$\mathcal{P}^{1,\delta_1+\delta_2} \widetilde{|\Lambda\rangle} = 0 , \qquad d = d_{12} . \tag{33}$$

The above conditions factorizes the submodule built on $v^1_{\delta_1+\delta_2}$. Thus, F^Λ_c is the UIR $L_\Lambda = F^\Lambda_c$.

Below $d < d_{12}$ there is no unitarity for $a_2 \neq 0$, except at the isolated point: $d_2 = n-2+\frac{1}{2}(a_2+\cdots+a_{n-1})$. At the latter point there is a singular vector $v^1_{\delta_2}$ which must be factored for unitarity. In addition, the previous singular vector is descendant of $v^1_{\delta_2}$ and the compact root singular vector $X^+_1 v_0$.

Further, for for $a_2 = 0$ the singular vectors $v^1_{\delta_1+\delta_2}$ and $v^1_{\delta_2}$ are descendants of the compact root singular vectors $X^+_1 v_0$ and $X^+_2 v_0$ which are factored out for $a_1 = a_2 = 0$. Thus, below we set also $a_2 = 0$ and there would be no obstacles for unitarity until the next reducibility points (coinciding due $a_2 = 0$): $d_2 = d_{13} = n-2+\frac{1}{2}(a_3+\cdots+a_{n-1})$. The singular vector for $d = d_{13}$ and $m = 1$ has weight $\delta_1+\delta_3 = \alpha_1+\alpha_2+2\alpha_3+\cdots 2\alpha_n$ and for $a_1 = 0$ it is a descendant of the compact root singular vector $X_1 v_0$ [70]. However, at $d_2 = d_{13}$ there is a subsingular vector which must be factored for unitarity. For $d < d_2 = d_{13}$ and $a_3 \neq 0$ the norm of that subsingular vector is negative, and there will be no unitarity except at some lower reducibility points.

For $d_{23} = n - \frac{5}{2} + \frac{1}{2}(a_3 + \cdots + a_{n-1})$ there is singular vector $v^1_{\delta_2+\delta_3}$ of weight $\delta_2 + \delta_3 = \alpha_2 + 2\alpha_3 + \cdots 2\alpha_n$ [70] which must be factored for unitarity. The previous subsingular vector is also factored out since it is descendant of $v^1_{\delta_2+\delta_3}$ and compact root singular vectors.

Further on, the Proof goes on similar lines. We list the points at which there are subsingular vectors - these happen when reducibility points coincide due the zero values of some a_i :

$$\begin{aligned}
& d_2 = d_{13} = n - 2 + \tfrac{1}{2}(a_3 + \cdots + a_{n-1}) \ , \quad a_1 = a_2 = 0 \ , \qquad (34)\\
& d_{23} = d_{14} = n - 5/2 + \tfrac{1}{2}(a_4 + \cdots + a_{n-1}) \ , \quad a_1 = a_2 = a_3 = 0, \ n > 3,\\
& d_3 = d_{24} = d_{15} = n - 3 + \tfrac{1}{2}(a_5 + \cdots + a_{n-1}) \ , \quad a_1 = a_2 = a_3 = a_4 = 0, \\
& \qquad n > 3,\\
& \ldots\\
& d_j = d_{1,2j-1} = d_{2,2j-2} = \cdots = d_{j-1,j+1} = n - j + \tfrac{1}{2}(a_{2j-1} + \cdots + a_{n-1}) \ ,\\
& \qquad a_1 = \cdots = a_{2j-2} = 0, \quad j < n,\\
& d_{j,j+1} = d_{1,2j} = d_{2,2j-1} = \cdots = d_{j-1,j+2} = n - j - \tfrac{1}{2} + \tfrac{1}{2}(a_{2j} + \cdots + a_{n-1}) \ ,\\
& \qquad a_1 = \cdots = a_{2j-1} = 0, \quad j < n-1.
\end{aligned}$$

Above it is understood that $a_j \equiv 0$ for $j \geq n$.
At the points of the subsingular vectors the associated singular vectors are factored out automatically. This happens also when the subsingular vectors are inside a continuous part of the unitarity spectrum. ■

The Proof above is not as explicit as we would like it to be, but due to the lack of space we postpone it to [74]. Below we give separately and explicitly the case $n = 3$.

Example: n=3. For $n = 3$ f-la (28) simplifies to:

$$\begin{array}{ccccccccc}
d_1 & > & d_{12} & > & d_2 & > & d_{23} & > & d_3 \\
 & & & \searrow & > \ d_{13} \ > & \nearrow & & &
\end{array}$$

The Theorem now reads:

$$\begin{aligned}
d &\geq 2 + \tfrac{1}{2}(a_1 + a_2) = d_1 \ , \quad a_1 \neq 0 \ , \qquad (35)\\
d &\geq \tfrac{3}{2} + \tfrac{1}{2}a_2 = d_{12} \ , \quad a_1 = 0, \ a_2 \neq 0 \ ,\\
d &= 1 + \tfrac{1}{2}a_2 = d_2 > d_{13} \ , \quad a_1 = 0, \ a_2 \neq 0 \ ,\\
d &\geq 1 = d_2 = d_{13} \ , \quad a_1 = a_2 = 0 \ ,\\
d &= \tfrac{1}{2} = d_{23} \ , \quad a_1 = a_2 = 0 \ ,\\
d &= 0 = d_3 \ , \quad a_1 = a_2 = 0 \ .
\end{aligned}$$

For $d > d_1$ there are no singular vectors and we have unitarity. At $d = d_1$ there is a singular vector:

$$v^1_{\delta_1} = \Big(a_1(a_1 + a_2 + 1)X_{\delta_1} - a_1 X_{\delta_3} X_{13} - (a_1 + a_2 + 1)X_{\delta_2} X^+_1 + + X_{\delta_3} X^+_2 X^+_1\Big) v_0 \tag{36}$$

which is given in PBW basis, where $X_{\delta_j} \in \mathcal{G}^+$ are the vectors corresponding to the weight vectors δ_j, X_{13} is the compact root vector for $\alpha_{13} = \alpha_1 + \alpha_2 = \delta_1 - \delta_3$. This singular vector is non-trivial for $a_1 \neq 0$ and must be eliminated to obtain an UIR. Below $d < d_1$ there is no unitarity for $a_1 \neq 0$. On the other hand for $a_1 = 0$ the singular vector $v^1_{\delta_1}$ is descendant of the compact root singular vector $X^+_1 v_0$ which is already factored out for $a_1 = 0$. Thus, below we discuss only the cases with $a_1 = 0$.
The singular vector at $d = d_{12}$ corresponding to the root $\delta_1 + \delta_2 = \alpha_1 + 2\alpha_2 + 2\alpha_3$ is:

$$\begin{aligned} v^1_{\delta_1+\delta_2} = \Big(& X_{\delta_3} X_{\delta_2} X^+_2 X^+_1 + \frac{1}{2} (X_{\delta_3})^2 (X^+_2)^2 X^+_1 - a_1 (X_{\delta_3})^2 X^+_2 X_{13} \\ & + 2(a_2+1)\, X_{\delta_3} X_{\delta_2} X_{13} - 2(a_1+a_2+1)\, X_{\delta_3} X_{\delta_1} X^+_2 \\ & + (a_1+1)(a_1+a_2+1)\, X_{\delta_1+\delta_3} X^+_2 + 4a_2 (a_1+a_2+1)\, X_{\delta_2} X_{\delta_1} \\ & + 2a_2 (a_1+a_2+1)\, (X_{\delta_2})^2 X^+_1 - \frac{1}{2} (a_1 + 2a_2 + 1)\, X_{\delta_2+\delta_3} X^+_2 X^+_1 \\ & - (-a_2 a_1 + a_1 + a_2 + 1)\, X_{\delta_2+\delta_3} X_{13} \\ & - 2(a_1+1)\, a_2 (a_1+a_2+1)\, X_{\delta_1+\delta_2} \Big) v_0 \end{aligned} \tag{37}$$

with norm:

$$16\,(2d - a_2 - 3)\left(a_1^2 + 2a_1 + 2d - a_2 - 2\right) a_2 (a_2+1)(a_1+a_2+1)(a_1+a_2+2)\,.$$

For $a_1 = 0,\ a_2 \neq 0$ it is non-trivial and gives rise to a invariant subspace which must be factored out for unitarity. For $d < d_{12}$ the vector (37) has negative norm and there is no unitarity for $a_2 \neq 0$, except at the isolated unitary point $d = 1 + \frac{1}{2}a_2 = d_2 > d_{13}$. At this point there is a singular vector v_{s2}, while the vector (37) is descendant of compact root singular vector $X_1 v_0$ and v_{s2}.
Further, we consider $a_1 = a_2 = 0$. Then the vector $v^1_{\delta_1+\delta_2}$ is descendant of compact root singular vectors $X^+_1 v_0$ and $X^+_2 v_0$, thus, there is no obstacle for unitarity for $1 < d$. The next reducibility points (coinciding here) are $d = d_{13} = d_2 = 1$. The singular vector for $d = d_2$ and $m = 1$ has weight $\delta_2 = \alpha_2 + \alpha_3$ and is given by:

$$v^1_{\delta_2} = \Big(a_2 X^+_2 X^+_3 - (a_2+1) X^+_3 X^+_2\Big) v_0 \tag{38}$$

For $a_2 = 0$ it is a descendant of the compact root singular vector $X_2^+ v_0$. The singular vector for $d = d_{13} = 1$ and $m = 1$ has weight $\delta_1 + \delta_3 = \alpha_1 + \alpha_2 + 2\alpha_3$ [70] :

$$\begin{aligned} v^1_{\delta_1+\delta_3} &= \Big(ha_1\, X_1^+ (X_3^+)^2 X_2^+ + a_1\, X_1^+ X_3^+ X_2^+ X_3^+ - \\ &\quad - h(a_1+1)(X_3^+)^2 X_2^+ X_1^+ - \\ &\quad - ha_1\, X_1^+ X_2^+ (X_3^+)^2 - (a_1+1) X_3^+ X_2^+ X_3^+ X_1^+ + \\ &\quad + h(a_1+1) X_2^+ (X_3^+)^2 X_1^+ \Big), \qquad (39) \\ &\qquad h = 1 + \tfrac{1}{2}(a_1 + a_2) \end{aligned}$$

The above vector is given in the simple root basis most appropriate for the case. For $a_1 = 0$ it is a descendant of the compact root singular vector $X_1^+ v_0$. However, there is a subsingular vector:

$$v_{ss} = (X_{\delta_1} X_{\delta_2} X_{\delta_3} - X_{\delta_3} X_{\delta_2} X_{\delta_1}) v_0 \qquad (40)$$

with norm: $16d(d-1)(2d-1)$. This must be factorized in order to obtain UR. Then for $\frac{1}{2} < d < 1$ there will be no unitarity due to the last norm.

Finally, at the next reducibility point: $d = d_{23} = \frac{1}{2}$ there is a singular vector of weight $\delta_2 + \delta_3 = \alpha_2 + 2\alpha_3$:

$$v^1_{\delta_2+\delta_3} = (2X_{\delta_2+\delta_3} - 4X_{\delta_2} X_{\delta_3} + X_{2\delta_3} X_2^+) v_0 \qquad (41)$$

It should be factored out to get unitarity. The subsingular vector (40) has zero norm for $d = \frac{1}{2}$ and furthermore it is descendant of $v^1_{\delta_2+\delta_3}$ and the compact root singular vector $X_2^+ v_0$. Finally, for $d < \frac{1}{2}$ there is no unitarity since then the norm of (41) is negative, except at the trivial isolated unitary point $d = 0 = a_1 = a_2$ of one-dimensional irrep. ■

4. Character formulae

4.1. Character formulae: generalities

In the beginning of this subsection we follow [73]. Let $\hat{\mathcal{G}}$ be a simple Lie algebra of rank ℓ with Cartan subalgebra $\hat{\mathcal{H}}$, root system $\hat{\Delta}$, simple root system $\hat{\pi}$. Let Γ, (resp. Γ_+), be the set of all integral, (resp. integral dominant), elements of $\hat{\mathcal{H}}^*$, i.e., $\lambda \in \hat{\mathcal{H}}^*$ such that $(\lambda, \alpha_i^\vee) \in \mathbb{Z}$, (resp. $\mathbb{Z}_+$), for all simple roots α_i, $(\alpha_i^\vee \equiv 2\alpha_i/(\alpha_i,\alpha_i))$. Let V be a lowest weight module with lowest weight Λ and lowest weight vector v_0. It has the following decomposition:

$$V = \bigoplus_{\mu\in\Gamma_+} V_\mu \,, \qquad V_\mu = \{u \in V \mid Hu = (\lambda+\mu)(H)u,\ \forall\, H \in \mathcal{H}\} \qquad (42)$$

(Note that $V_0 = \mathbb{C}v_0$.) Let $E(\mathcal{H}^*)$ be the associative abelian algebra consisting of the series $\sum_{\mu\in\mathcal{H}^*} c_\mu e(\mu)$, where $c_\mu \in \mathbb{C}$, $c_\mu = 0$ for μ outside the union of a finite number of sets of the form $D(\lambda) = \{\mu \in \mathcal{H}^* | \mu \geq \lambda\}$, using some ordering of $\mathcal{H}^*$, e.g., the lexicographic one; the formal exponents $e(\mu)$ have the properties: $e(0) = 1$, $e(\mu)e(\nu) = e(\mu + \nu)$.

Then the (formal) character of V is defined by:

$$ch_0\ V \ = \ \sum_{\mu\in\Gamma_+} (\dim\ V_\mu)\ e(\Lambda+\mu) \ = \ e(\Lambda) \sum_{\mu\in\Gamma_+} (\dim\ V_\mu)\ e(\mu) \qquad (43)$$

(We shall use subscript '0' for the even case.)

For a Verma module, i.e., $V = V^\Lambda$ one has $\dim\ V_\mu = P(\mu)$, where $P(\mu)$ is a generalized partition function, $P(\mu) = \#$ of ways μ can be presented as a sum of positive roots β, each root taken with its multiplicity $\dim \mathcal{G}_\beta$ $(= 1$ here), $P(0) \equiv 1$. Thus, the character formula for Verma modules is:

$$ch_0\ V^\Lambda \ = \ e(\Lambda) \sum_{\mu\in\Gamma_+} P(\mu)e(\mu) \ = \ e(\Lambda) \prod_{\alpha\in\Delta^+} (1-e(\alpha))^{-1}. \qquad (44)$$

Further we recall the standard reflections in $\hat{\mathcal{H}}^*$:

$$s_\alpha(\lambda) \ = \ \lambda - (\lambda, \alpha^\vee)\alpha\,, \quad \lambda \in \hat{\mathcal{H}}^*\,, \quad \alpha \in \hat{\Delta}. \qquad (45)$$

The Weyl group W is generated by the simple reflections $s_i \equiv s_{\alpha_i}$, $\alpha_i \in \hat{\pi}$. Thus every element $w \in W$ can be written as the product of simple reflections. It is said that w is written in a reduced form if it is written with the minimal possible number of simple reflections; the number of reflections of a reduced form of w is called the length of w, denoted by $\ell(w)$.

The Weyl character formula for the finite-dimensional irreducible LWM L_Λ over $\hat{\mathcal{G}}$, i.e., when $\Lambda \in -\Gamma_+$, has the form:

$$ch_0\ L_\Lambda \ = \ \sum_{w\in W} (-1)^{\ell(w)}\ ch_0\ V^{w\cdot\Lambda}\,, \quad \Lambda \in -\Gamma_+ \qquad (46)$$

where the dot $\cdot$ action is defined by $w \cdot \lambda = w(\lambda - \rho) + \rho$. For future reference we note:

$$s_\alpha \cdot \Lambda \ = \ \Lambda + n_\alpha \alpha \qquad (47)$$

where

$$n_\alpha \ = \ n_\alpha(\Lambda) \ \doteq \ (\rho - \Lambda, \alpha^\vee) \ = \ (\rho - \Lambda)(H_\alpha)\,, \quad \alpha \in \Delta^+. \qquad (48)$$

In the case of basic classical Lie superalgebras the first character formulae were given by Kac [63, 71].[1] For all such superalgebras – except

[1] Kac considers highest weight modules but his results are immediately transferable to lowest weight modules.

$osp(1/2n)$ – the character formula for Verma modules is [63, 71]:

$$ch\ V^\Lambda \ = e(\Lambda) \left(\prod_{\alpha\in\Delta^+_{\bar 0}} (1-e(\alpha))^{-1} \right) \left(\prod_{\alpha\in\Delta^+_{\bar 1}} (1+e(\alpha)) \right) . \tag{49}$$

We are however interested exactly in the $osp(1/2n)$ when the Verma module character formula is:

$$ch\ V^\Lambda \ = e(\Lambda) \left(\prod_{\alpha\in\bar\Delta^+} (1-e(\alpha))^{-1} \right) \tag{50}$$

Naturally, the character formula for the finite-dimensional irreducible LWM L_Λ is again (46) using the Weyl group W_n of B_n.

4.2. Multiplets

A Verma module V^Λ may be reducible w.r.t. to many positive roots, and thus there maybe many Verma modules isomorphic to its submodules. They themselves may be reducible, and so on.

One main ingredient of the approach of [66] is as follows. We group the (reducible) Verma modules with the same Casimirs in sets called *multiplets* [69]. The multiplet corresponding to fixed values of the Casimirs may be depicted as a connected graph, the vertices of which correspond to the reducible Verma modules and the lines between the vertices correspond to embeddings between them. The explicit parametrization of the multiplets and of their Verma modules is important for understanding of the situation.

If a Verma module V^Λ is reducible w.r.t. to all simple roots (and thus w.r.t. all positive roots), i.e., $m_k \in I\!\!N$ for all k, then the irreducible submodules are isomorphic to the finite-dimensional irreps of $\mathcal{G}^{I\!\!C}$ [66]. (Actually, this is a condition only for m_n since $m_k \in I\!\!N$ for $k = 1,\ldots,n-1$.) In these cases we have the *main multiplets* which are isomorphic to the Weyl group of $\mathcal{G}^{I\!\!C}$ [66].

In the cases of non-dominant weight Λ the character formula for the irreducible LWM is [72] :

$$ch\ L_\Lambda \ = \ \sum_{\substack{w\in W \\ w\leq w_\Lambda}} (-1)^{\ell(w_\Lambda w)}\ P_{w,w_\Lambda}(1)\ ch\ V^{w\cdot(w_\Lambda^{-1}\cdot\Lambda)}\ , \quad \Lambda\in\Gamma \tag{51}$$

where $P_{y,w}(u)$ are the Kazhdan–Lusztig polynomials $y, w \in W$ [72] (for an easier exposition see [68]), w_Λ is a unique element of W with minimal length such that the signature of $\Lambda_0 \ = \ w_\Lambda^{-1}\cdot\Lambda$ is anti-dominant or semi-anti-dominant:

$$\chi_0 \ = \ (m'_1,\ldots,m'_n), \qquad m'_k \ = \ 1-\Lambda_0(H_k) \ \in Z\!\!\!Z_-\ . \tag{52}$$

Note that $P_{y,w}(1) \in \mathbb{N}$ for $y \leq w$.

When Λ_0 is semi-anti-dominant, i.e., at least one $m'_k = 0$, then in fact W is replaced by a reduced Weyl group W_R.

Most often the value of $P_{y,w}(1)$ is equal to 1 (as in the character formula for the finite-dimensional irreps), while the cases $P_{y,w}(1) > 1$ are related to the appearance of subsingular vectors, though the situation is more subtle, see [68].

It is interesting to see how the reducible points relevant for unitarity fit in the multiplets. In the case of d_{ij} (26) and using (13) we have:

$$m_n(d_{ij}) \;=\; 1 - 2m_j - \cdots - 2m_{n-1} - m_i - \cdots - m_{j-1} \,. \tag{53}$$

In the case of d_i (27) we have:

$$m_n(d_i) \;=\; 1 - 2m_i - \cdots - 2m_{n-1} \,. \tag{54}$$

As expected the weights related to positive energy d are not dominant ($m_n(d_{ij}) \in \mathbb{Z}_-$, $m_n(d_i) \in -\mathbb{N}$, $(i < n)$), since the positive energy UIRs are infinite-dimensional. (Naturally, $m_n(d_n) = 1$ falls out of the picture since $d_n < 0$.)

Thus, the Verma modules with weights related to positive energy would be somewhere in the main multiplet (or in a reduction of the main multiplet), and the first task for calculating the character is to find the w_Λ in the character formula (51). This we do in the next subsection in the case $n = 3$.

4.3. The case n=3

In order to illustrate what the main ideas we consider the first non-trivial example $n = 3$, i.e., $osp(1/6)$ actually using B_3. The Weyl group W_n of B_n has $2^n n!$ elements, i.e., 48 for B_3. Let $S = (s_1, s_2, s_3)$, $s_i \equiv s_{\alpha_i}$, be the simple reflections. They fulfill the following relations:

$$s_1^2 = s_2^2 = s_3^2 = e,\ (s_1 s_2)^3 = e,\ (s_2 s_3)^4 = e,\ s_1 s_3 \;=\; s_3 s_1 \,, \tag{55}$$

e being the identity of W_3. The 48 elements may be listed as:

$$
\begin{aligned}
& e\,,\ s_1\,,\ s_2\,,\ s_3 \\
& s_1s_2\,,\ s_1s_3\,,\ s_2s_1\,,\ s_2s_3\,,\ s_3s_2\,, \\
& s_1s_2s_1\,,\ s_1s_2s_3\,,\ s_1s_3s_2\,,\ s_2s_1s_3\,,\ s_2s_3s_2\,,\ s_3s_2s_1\,,\ s_3s_2s_3\,, \\
& s_1s_2s_1s_3\,,\ s_1s_2s_3s_2\,,\ s_1s_3s_2s_1\,,\ s_1s_3s_2s_3\,, \\
& s_2s_3s_2s_1\,,\ s_2s_1s_3s_2\,,\ s_3s_2s_3s_1\,,\ s_3s_2s_3s_2\,, \\
& s_1s_2s_3s_2s_1\,,\ s_1s_3s_2s_1s_3\,,\ s_1s_2s_1s_3s_2\,,\ s_1s_3s_2s_3s_2\,, \\
& s_2s_1s_3s_2s_1\,,\ s_2s_1s_3s_2s_3\,,\ s_3s_2s_3s_1s_2\,,\ s_3s_2s_3s_2s_1\,, \\
& s_1s_3s_2s_3s_2s_1\,,\ s_1s_3s_2s_1s_3s_2\,,\ s_1s_2s_1s_3s_2s_1\,,\ s_2s_1s_3s_2s_1s_3\,, \\
& s_2s_1s_3s_2s_3s_2\,,\ s_3s_2s_3s_1s_2s_1\,,\ s_3s_2s_3s_1s_2s_3\,, \\
& s_2s_1s_3s_2s_3s_2s_1\,,\ s_2s_1s_3s_2s_3s_1s_2\,,\ s_3s_2s_1s_2s_3s_2s_1\,, \\
& s_3s_2s_3s_1s_2s_1s_3\,,\ s_3s_2s_3s_1s_2s_3s_2\,, \\
& s_2s_3s_2s_1s_2s_3s_2s_1\,,\ s_3s_2s_1s_3s_2s_3s_2s_1\,,\ s_3s_2s_1s_3s_2s_3s_1s_2\,,
\end{aligned}
\tag{56}
$$

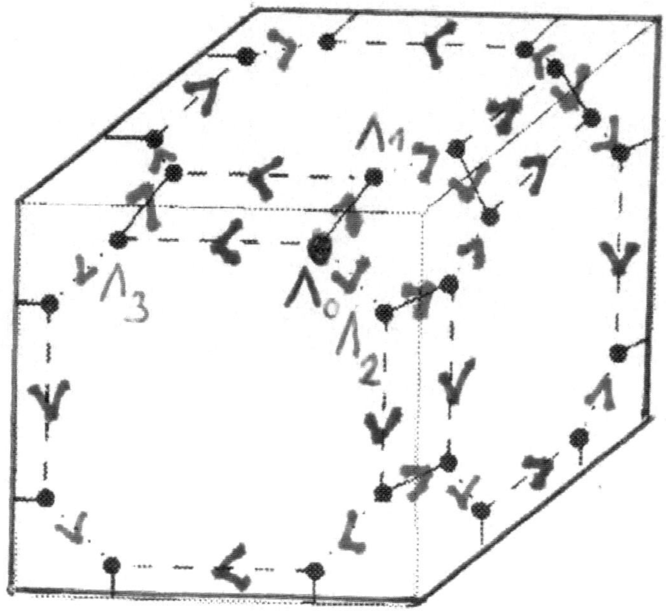

This Weyl group may be pictorially represented on a cube as in the figure, where we have given only the simple root reflections, namely, continuous (red) arrows represent action of reflection s_1, dashed (blue) arrows represent action of reflection s_2, dotted (green) arrows represent action of reflection s_3. Each face of the cube contains eight elements related by blue and green arrows representing the Weyl group of B_2 generated by s_2 and s_3. The figure contains also eight sextets (around the eight corners of the cube). Each sextet is related by red and green arrows representing the Weyl group of A_2 generated by s_1 and s_2. Finally there are 12 quartets (straddling the edges of the cube). Each quartet is formed by red and blue arrows representing the Weyl group of $A_1 \times A_1$ generated by the commuting reflections s_1 and s_3.

We use the same diagram to depict the main multiplets containing the Verma modules V^{Λ_0} which contain (as factor module) the finite-dimensional irreps of B_3, i.e., with dominant weights Λ_0, i.e., with Dynkin labels (m_1, m_2, m_3), $m_k \in I\!N$. We may do this since these multiplets are isomorphic to the Weyl group, W_3 in our case. On the picture we have indicated the modules, Λ_0 and $\Lambda_k = s_k \cdot \Lambda_0$, $k = 1, 2, 3$. The mentioned isomorphism is fixed by assigning to Λ_0 the identity element e of W_3, and to Λ_k the reflections s_k.

The character formula for the Verma modules in our case is given explicitly by:

$$
\begin{aligned}
ch\ V^{\Lambda} &= \frac{e(\Lambda)}{(1-t_1)(1-t_2)(1-t_1t_2)} \times \\
&\times \frac{1}{(1-t_3)(1-t_2t_3)(1-t_1t_2t_3)(1-t_2t_3^2)(1-t_1t_2t_3^2)(1-t_1t_2^2t_3^2)}
\end{aligned}
\tag{57}
$$

where $t_j \equiv e(\alpha_j)$.

Now we give the character formulae of the five boundary or isolated unitarity cases. Below we shall denote the signature of the dominant weight Λ_0 which determines the main multiplet by (m'_1, m'_2, m'_3), $m'_k \in I\!N$, using primes to distinguish from the signatures of the weights we are interested. We shall use also reductions of the main multiplet when the weights are semi-dominant, i.e., when some $m'_k = 0$.

• In the case of $d = d_1 = 2 + \frac{1}{2}(a_1 + a_2)$ there are twelve members of the multiplet which is a submultiplet of a main multiplet. (Remember that that $m_1 > 1$ since $a_1 \neq 0$.) They are grouped into two standard $sl(3)$ submultiplets of six members. The first submultiplet starts from $V^{\Lambda_0^{d_1}}$, where $\Lambda_0^{d_1} = w \cdot \Lambda_0$, $w = w_{\Lambda_0^{d_1}} = s_2 s_1 s_3 s_2 s_3$, with signature:

$$
\Lambda_0^{d_1} \ :\ (m_1, m_2, m'_3 = 1 - 2m_{12})\ , \quad m_1, m_2 \in I\!N\ , \quad m_{12} \equiv m_1 + m_2\ . \tag{58}
$$

The other submultiplet starts from $V^{\Lambda'_0}$ with $\Lambda'_0 = \Lambda_0^{d_1} + \delta_1 = \Lambda_0^{d_1} + \alpha_1 + \alpha_2 + \alpha_3$, with signature: $\Lambda'_0\ :\ (m_1 - 1, m_2, m'_3 = 1 - 2m_{12})$, $m_1 > 1$.

The character formula is (51) with $w_\Lambda = w_{\Lambda_0^{d_1}}$:

$$\mathrm{ch}\,\Lambda_0^{d_1} = \frac{e(\Lambda_0^{d_1})}{(1-t_3)(1-t_2t_3)(1-t_1t_2t_3)(1-t_2t_3^2)(1-t_1t_2t_3^2)(1-t_1t_2^2t_3^2)}$$
$$\times \{\ \mathrm{ch}\,\Lambda_{m_1,m_2}(t_1,t_2) \ - \ t_1t_2t_3\,\mathrm{ch}\,\Lambda_{m_1-1,m_2}(t_1,t_2)\ \}\ , \quad m_1 > 1 \qquad (59)$$

where $\mathrm{ch}\,\Lambda_{m_1,m_2}(t_1,t_2)$ is the normalized character of the finite-dimensional $sl(3)$ irrep with Dynkin labels (m_1,m_2) (and dimension $m_1m_2(m_1+m_2)/2$):

$$\mathrm{ch}\ \Lambda_{m_1,m_2}(t_1,t_2) \ = \ \frac{1-t_1^{m_1}-t_2^{m_2}+t_1^{m_1}t_2^{m_{12}}+t_1^{m_{12}}t_2^{m_2}-t_1^{m_{12}}t_2^{m_{12}}}{(1-t_1)(1-t_2)(1-t_1t_2)} \qquad (60)$$

Naturally, the latter formula is a polynomial in t_1,t_2 , e.g., $\mathrm{ch}\ \Lambda_{1,1}(t_1,t_2) = 1$. Note that (59) trivializes for $m_1 = 1$ since the second term disappears by the formal substitution: $\mathrm{ch}\ \Lambda_{0,m_2}(t_1,t_2) = 0$.

• In the case of $d = d_{12} = \frac{1}{2}(3+a_2)$ which is relevant for unitarity, i.e., $m_1 = 1$, there are again twelve members of the multiplet. The corresponding signature is:

$$\Lambda_0^{d_{12}}\ :\ (1, m_2, m_3' = -2m_2)\ , \qquad m_2 \in I\!N\ . \qquad (61)$$

The multiplet is submultiplet of a reduced multiplet with 24 members obtained from a main multiplet for $m_3' = 0$. As above our multiplet consists of two standard $sl(3)$ submultiplets of six members. The first submultiplet starts from $V^{\Lambda_0^{d_{12}}}$, where $\Lambda_0^{d_{12}} = w \cdot \Lambda_0$, $w = w_{\Lambda_0^{d_{12}}} = s_3s_2s_1$. The other submultiplet starts from $V^{\Lambda_0'}$ with $\Lambda_0' = \Lambda_0^{d_{12}} + m_2(\alpha_1 + 2\alpha_2 + 2\alpha_3) = \Lambda_0^{d_{12}} + m_2(\delta_1+\delta_2)$ with signature: $\Lambda_0'\ :\ (1, m_2-1, -2m_2)$. The character formula is (51), with $W \mapsto W_R$, (where W_R is a reduced 24-member Weyl group) and with $w_\Lambda = w_{\Lambda_0^{d_{12}}}$:

$$\mathrm{ch}\,\Lambda_0^{d_{12}} = \frac{e(\Lambda_0^{d_{12}})}{(1-t_3)(1-t_2t_3)(1-t_1t_2t_3)(1-t_2t_3^2)(1-t_1t_2t_3^2)(1-t_1t_2^2t_3^2)}$$
$$\times\{\ \mathrm{ch}\,\Lambda_{1,m_2}(t_1,t_2) \ - \ (t_1t_2^2t_3^2)^{m_2}\,\mathrm{ch}\,\Lambda_{1,m_2-1}(t_1,t_2)\ \}\ , \quad m_2 > 1 \qquad (62)$$

where $\mathrm{ch}\,\Lambda_{m_1,m_2}$ are the $sl(3)$ characters defined in (60).

• In the case of $d = d_2 = 1 + \frac{1}{2}a_2 \geq d_{13}$, i.e., $m_3' = 1 - 2m_2$, the corresponding signature is:

$$\Lambda_0^{d_2}\ :\ (m_1, m_2, m_3' = 1-2m_2)\ . \qquad (63)$$

We should consider two subcases

$$1+m_1-m_2>0 \qquad \text{or} \qquad 1+m_1-m_2\leq 0$$

We start with the *first subcase* which is relevant when $d=d_2=d_{13}=1$ and $a_1=a_2=0$, then $m_1=m_2=1$, and the signature is:

$$\Lambda_0^{d_2=d_{13}} \ : \ (1,1,-1) \ . \tag{64}$$

Our multiplet is a submultiplet of a 12-member reduced multiplet obtained when the signature of Λ_0 is $(m_1',m_2',m_3') = (1,0,1)$, and then $\Lambda_0^{d_2=d_{13}}$ is a submodule with signature (64). Thus, we have $\Lambda_0^{d_2=d_{13}} = s_3\cdot\Lambda_0$, i.e., $w_{\Lambda_0^{d_2=d_{13}}} = s_3$.

Explicitly, our 12-member multiplet has two $sl(3)$ submultiplets. First we take into account a $sl(3)$ sextet starting from $\Lambda_0^{d_2=d_{13}}$ with parameters $(1,1)$. Then there is a $sl(3)$ sextet starting from $\Lambda_0^{d_2=d_{13}}+\alpha_1+2\alpha_2+3\alpha_3$ with parameters $(1,1)$. Note that that $\alpha_1+2\alpha_2+3\alpha_3=\delta_1+\delta_2+\delta_3$ is the weight of the subsingular vector (40).

The character formula is (51), with $W\mapsto W_R$, (where W_R is a reduced 12-member Weyl group) and $w_\Lambda = s_3$:

$$\begin{aligned}
\mathrm{ch}\,\Lambda_0^{d_2=d_{13}} &= \\
&= \frac{e(\Lambda_0^{d_2=d_{13}})}{(1-t_3)(1-t_2t_3)(1-t_1t_2t_3)(1-t_2t_3^2)(1-t_1t_2t_3^2)(1-t_1t_2^2t_3^2)} \times \\
&\times \{\ 1 \ - \ t_1t_2^2t_3^3\ \}
\end{aligned} \tag{65}$$

• In the case of $d=d_2=1+\frac{1}{2}a_2>d_{13}=1$, i.e., $m_1=1$, $m_2=1+a_2>1$, thus, this is the subcase $1+m_1-m_2=m_{13}\leq 0$. The multiplet has 24 members for $m_2>2$ $(m_{13}<0)$ and starts with $\Lambda_0^{'d_2} = \ = s_3\,s_2\,s_1\cdot\Lambda_0$, with signatures:

$$\begin{aligned}
&\Lambda_0 \ : \ (m_2-2,1,1) \ , \\
&\Lambda_0^{'d_2} \ : \ (1,m_2,m_3'=1-2m_2) \ , \qquad m_2\in 1+I\!N \ .
\end{aligned} \tag{66}$$

It has four $sl(3)$ submultiplets. First we take into account a $sl(3)$ sextet starting from $\Lambda_0^{'d_2}$ with parameters $(1,m_2)$. Then there is a $sl(3)$ sextet starting from $\Lambda_0^{'d_2}+\alpha_{23}$ with parameters $(2,m_2-1)$. Then there is a $sl(3)$ sextet with parameters $(2,m_2-2)$ starting from a Verma module $V^{\Lambda''}$, $\Lambda''=\Lambda_0^{'d_2}+\alpha_1+3\alpha_{23}$. Finally, there is a $sl(3)$ sextet with parameters $(1,m_2-2)$, starting from a Verma module $V^{\Lambda'''}$, $\Lambda'''=\Lambda_0^{'d_2}+2(\alpha_1+2\alpha_2+$

$2\alpha_3$).
We have the *Conjecture* that the character formula is (51) and $w_\Lambda = s_3\, s_2\, s_1$:

$$\mathrm{ch}\,\Lambda_0^{'d_2} = \frac{e(\Lambda_0^{'d_2})}{(1-t_3)(1-t_2t_3)(1-t_1t_2t_3)(1-t_2t_3^2)(1-t_1t_2t_3^2)(1-t_1t_2^2t_3^2)}$$
$$\times \{\ \mathrm{ch}\,\Lambda_{1,m_2}(t_1,t_2)\ -\ t_2t_3\,\mathrm{ch}\,\Lambda_{2,m_2-1}(t_1,t_2)\ +$$
$$+\ t_1t_2^3t_3^3\,\mathrm{ch}\,\Lambda_{2,m_2-2}(t_1,t_2)\ -\ t_1^2t_2^4t_3^4\,\mathrm{ch}\,\Lambda_{1,m_2-2}(t_1,t_2)\ \} \qquad (67)$$

When $m_2 = 2$ $(a_2 = 1,\ m_{13} = 0)$ the weight Λ_0 is semi dominant, the main multiplet reduces to 24 members, our multiplet reduces to only 12 members, consisting of the first two $sl(3)$ submultiplets mentioned above. The character formula takes this into account by construction since for $m_2 = 2$ the terms in the 2nd row are automatically zero (due to the fact that the $sl(3)$ character formula gives zero: $\mathrm{ch}\,\Lambda_{1,0}(t_1,t_2) = 0$).

• In the case of $d = d_{23} = \frac{1}{2}$, $a_1 = a_2 = 0$, i.e., $m_1 = m_2 = 1$, and the signature is:

$$\Lambda_0^{d_{23}}\ :\ (1,1,0)\ . \qquad (68)$$

This is in fact a multiplet with 24 members which is reduction of the main multiplet starting with the semi dominant weight (68).

The multiplet consists of four $sl(3)$ submultiplets. First there is a $sl(3)$ sextet starting from $\Lambda_0^{d_{23}}$ with parameters $(1,1)$. Then a $sl(3)$ sextet starting from $\Lambda_0^{d_{23}} + \alpha_2 + 2\alpha_3$ with parameters $(2,1)$. Then a $sl(3)$ sextet starting from $\Lambda_0^{d_{23}} + \alpha_1 + 2\alpha_2 + 4\alpha_3$ with parameters $(1,2)$. Then a $sl(3)$ sextet starting from $\Lambda_0^{d_{23}} + 2\alpha_1 + 4\alpha_2 + 6\alpha_3$ with parameters $(1,1)$.

The character formula is (51), however, with $W \mapsto W_R$, where W_R is the reduced 24-member Weyl group, (generated by $s_1, s_2, s_3s_2s_3$)

and $w_\Lambda = 1$:

$$\begin{aligned}
\mathrm{ch}\,\Lambda_0^{d_{23}} &= \frac{e(\Lambda_0^{d_{23}})}{(1-t_3)(1-t_2t_3)(1-t_1t_2t_3)(1-t_2t_3^2)(1-t_1t_2t_3^2)(1-t_1t_2^2t_3^2)} \\
&\quad \times \{\ 1\ -\ t_1t_2^2t_3^4\,\mathrm{ch}\,\Lambda_{1,2}(t_1,t_2)\ + \\
&\quad +\ t_2t_3^2\,\mathrm{ch}\,\Lambda_{2,1}(t_1,t_2)\ -\ t_1^2t_2^4t_3^6\ \}\ = \\
&= \frac{e(\Lambda_0^{d_{23}})}{(1-t_3)(1-t_2t_3)(1-t_1t_2t_3)(1-t_2t_3^2)(1-t_1t_2t_3^2)(1-t_1t_2^2t_3^2)} \\
&\quad \times \{\ 1\ -\ t_1t_2^2t_3^4\,(1+t_2+t_1t_2)\ + \\
&\quad +\ t_2t_3^2\,(1+t_1+t_1t_2)\ -\ t_1^2t_2^4t_3^6\ \}
\end{aligned} \tag{69}$$

Acknowledgements

VKD is supported in part by Bulgarian NSF Grant DFNI T02/6. I. Salom is supported in part by the Serbian Ministry of Science and Technological Development under grant number OI 171031.

References

[1] W. Nahm, Nucl. Phys. **B135**, 149 (1978).

[2] R. Haag, J.T. Lopuszanski and M. Sohnius, Nucl. Phys. **B88**, 257 (1975).

[3] M. Flato and C. Fronsdal, Lett. Math. Phys. **8**, 159 (1984).

[4] V.K. Dobrev and V.B. Petkova, Lett. Math. Phys. **9**, 287 (1985).

[5] V.K. Dobrev and V.B. Petkova, Phys. Lett. **162B**, 127 (1985).

[6] V.K. Dobrev and V.B. Petkova, Fortschr. d. Phys. **35**, 537 (1987).

[7] V.K. Dobrev and V.B. Petkova, Proceedings, eds. A.O. Barut and H.D. Doebner, Lecture Notes in Physics, Vol. 261 (Springer-Verlag, Berlin, 1986) p. 291 and p. 300.

[8] S. Minwalla, Adv. Theor. Math. Phys. **2**, 781 (1998).

[9] V.K. Dobrev, J. Phys. **A35** (2002) 7079; hep-th/0201076.

[10] P.K. Townsend, P-brane democracy, PASCOS/Hopkins 0271-286 (1995), hep-th/9507048; Four lectures on M theory, in: *Trieste 1996, High energy physics and cosmology* 385-438, hep-th/9612121; Nucl. Phys. Proc. Suppl. **68**, 11-16, (1998), hep-th/9708034; J.P. Gauntlett, G.W. Gibbons, C.M. Hull and P.K. Townsend, Commun. Math. Phys. **216**, 431-459, (2001), hep-th/0001024.

[11] R. D'Auria, S. Ferrara, M.A. Lledo and V.S. Varadarajan, J. Geom. Phys. **40** 101-128, (2001), hep-th/0010124; R. D'Auria, S. Ferrara and M.A. Lledo, Lett. Math. Phys. **57**, 123-133, (2001), hep-th/0102060; M.A. Lledo and V.S. Varadarajan, Spinor algebras and extended superconformal algebras. Proc. 2nd International Symposium on Quantum Theory and Symmetries, Cracow, Poland, 18-21 Jul 2001, hep-th/0111105; S. Ferrara and M.A. Lledo, Rev. Math. Phys. **14**, 519-530, (2002), hep-th/0112177.

[12] V.K. Dobrev and R.B. Zhang, Positive Energy Unitary Irreducible Representations of the Superalgebras $osp(1|2n, R)$, Phys. Atom. Nuclei, **68** (2005) 1660-1669; hep-th/0402039.

[13] V.K. Dobrev, A.M. Miteva, R.B. Zhang and B.S. Zlatev, On the Unitarity of D=9,10,11 Conformal Supersymmetry, Czech. J. Phys. **54** (2004) 1249-1256

[14] N. Beisert, Phys. Rept. **405**, 1-202 (2005); N. Beisert, H. Elvang, D. Z. Freedman, M. Kiermaier, A.Morales, S. Stieberger, Phys. Lett. **B694**, 265-271 (2010)

[15] B. Eden, C. Jarczak and E. Sokatchev, Nucl. Phys. **B712**, 157-195 (2005); J. Henn, C. Jarczak, E. Sokatchev, Nucl. Phys. **B730** 191-209 (2005).

[16] M. Bianchi, Fortsch. Phys. **53**, 665-691 (2005); M. Bianchi, P.J. Heslop, F. Riccioni, JHEP 0508:088 (2005); M. Bianchi, S. Kovacs, G. Rossi, Lect. Notes in Physics, v. 737 (2008) pp. 303-470;
S. Ananth, S. Kovacs & S. Parikh, JHEP 1205:096 (2012).

[17] C. Carmeli, G. Cassinelli, A. Toigo, V.S. Varadarajan, Comm. Math. Phys. **263** (2006) 217; C. Carmeli, G. Cassinelli, A. Toigo, in: Lie Theory and Its Applications in Physics VI, (Heron Press, Sofia, 2006) p. 269; V.S. Varadarajan, Unitary representations of super Lie groups, Lectures given in Oporto, Portugal, July 2023, 2006; R. Fioresi, M. A. Lledo, V. S. Varadarajan, J. Math. Phys. **48** (2007) 105017.

[18] A. Barabanschikov, L. Grant, L.L. Huang, S. Raju, JHEP 0601 (2006) 160; J. Kinney, J. Maldacena, Sh. Minwalla, S. Raju, An index for 4 dimensional super conformal theories, Commun. Math. Phys. **275**, 209-254 (2007); J. Bhattacharya, S. Bhattacharyya, S. Minwalla and S. Raju, JHEP 0802 (2008) 064.

[19] L. Genovese, Ya.S. Stanev, Nucl. Phys. **B721** (2005) 212;
M. D'Alessandro and L. Genovese, Nucl. Phys. **B732** (2006) 64.

[20] Gu. Milanesi and M. O'Loughlin, JHEP 09 (2005) 008;
E. Gava, G. Milanesi, K.S. Narain, M. O'Loughlin, JHEP 05 (2007) 030.

[21] R.R. Metsaev, Phys. Rev. **D71** (2005) 085017; Phys. Lett. **B636**, 227-233 (2006); JHEP 1201 (2012) 064; JHEP 1206 (2012) 117; Conformal totally symmetric arbitrary spin fermionic fields, arXiv:1211.4498.

[22] J. Terning, Modern supersymmetry: Dynamics and duality, International Series of Monographs on Physics # 132, (Oxford University Press, 2005) 336 pages;
J. Galloway, J. McRaven and J. Terning, Phys. Rev. **D80** (2009) 105017.

[23] Yu. Nakayama, Nucl. Phys. **B755**, 295-312 (2006); Phys. Rev. **D76**, 105009 (2007); JHEP 0810 (2008) 083;
M. Ibe, Y. Nakayama, T.T. Yanagida, Phys. Lett. **B668** (2008) 28.

[24] F.A. Dolan, J. Math. Phys. **47**, 062303 (2006); Nucl. Phys. **B790** (2008) 432; M. Bianchi, F.A. Dolan, P.J. Heslop & H. Osborn, Nucl. Phys. **B767**, 163-226, (2007).

[25] V.K. Dobrev, Phys. Part. Nucl. **38** (5) (2007) 564-609; Czech. J. Phys. **56** (2006) 1131-1136; Fortschr. Phys. **57**, 542545 (2009); Nucl. Phys. **B854** (2012) 878-893; Phys. Part. Nucl. **43** (2012) 616-620; J. Phys. **A46** (2013) 405202.

[26] M. Berkooz, D. Reichmann, J. Simon, JHEP 0701 (2007) 048;
O. Aharony, L. Berdichevsky, M. Berkooz, Y. Hochberg, D. Robles-Llana, Phys. Rev. **D81** (2010) 085006.

[27] T.A. Ryttov & F. Sannino, Phys. Rev. **D76**, 105004 (2007); Phys. Rev. **D78**, 065001 (2008); Int. J. Mod. Phys. **A25** (24) (2010) 4603; M. Jarvinen, F. Sannino, JHEP 1005 (2010) 041; F. Sannino, Int. J. Mod. Phys. **A25**, 5145-5161 (2010); T.A. Ryttov, R. Shrock, Phys. Rev. **D83**, 056011 (2011); Phys. Rev. **D85**, 076009 (2012); T.A. Ryttov, Phys. Rev. **D90** (2014) 056007.

[28] H. Murayama, Ya. Nomura, D. Poland, Phys. Rev. **D77**, 015005 (2008); D. Poland, JHEP 0911 (2009) 049; D. Poland, D. Simmons-Duffin, JHEP 1005 (2010) 079; JHEP 1105 (2011) 017; A.L. Fitzpatrick, J. Kaplan, Z.U. Khandker, D.L. Li, D. Poland, D. Simmons-Duffin, JHEP 1408 (2014) 129.

[29] L. Baulieu, G. Bossard , JHEP 0802 (2008) 075;
G. Bossard, P.S. Howe, K.S. Stelle, P. Vanhove, Class. Quant. Grav. 28:215005, (2011); G. Bossard, P.S. Howe, K.S. Stelle, JHEP 1307 (2013) 117,

[30] I. Heckenberger, F. Spill, A. Torrielli, H. Yamane, Drinfeld second realization of the quantum affine superalgebras of D(1)(2,1:x) via the Weyl groupoid, Publ. Res. Inst. Math. Sci. Kyoto B8 (2008) 171; I. Heckenberger & H. Yamane, Math. Z. **259**, 255-276 (2008).

[31] A. Solovyov, JHEP 0804 (2008) 013.

[32] A.D. Shapere and Y. Tachikawa, JHEP 09 (2008) 109; D. Green, Z. Komargodski, N. Seiberg, Yu. Tachikawa, B. Wecht, JHEP 06 (2010) 106.

[33] S. Lievens, N.I. Stoilova & J. Van der Jeugt, Commun. Math. Phys. 281, 805-826 (2008); J. Generalized Lie Theory and Appl., **2** (2008) 206.

[34] D.D. Dietrich, Phys. Rev. **D80**, 065032 (2009); Phys. Rev. **D82**, 065007 (2010).

[35] O. Antipin, K. Tuominen, Resizing the Conformal Window: A beta function Ansatz. Phys. Rev. **D81**, 076011 (2010); Mod. Phys. Lett. **A26**, 2227-2246 (2011).

[36] K. Yonekura, JHEP 1009:049 (2010); JHEP 1401 (2014) 142.

[37] T. Horigane, Y. Kazama, Phys. Rev. **D81** (2010) 045004.

[38] A. Babichenko, B. Stefanski, K. Zarembo, JHEP 1003 (2010) 058.

[39] N. Gromov, V. Kazakov, Z. Tsuboi, $PSU(2,2|4)$ character of quasiclassical AdS/CFT, JHEP 1007 (2010) 097.

[40] H. Knuth, Int. J. Mod. Phys. **A26** (2011) 2007.

[41] A. Gadde, L. Rastelli, S.S. Razamat and W.Yan, On the Superconformal Index of N=1 IR Fixed Points: A Holographic Check, JHEP 1103:041, (2011); Commun. Math. Phys. **319** (2013) 147; A. Gadde, E. Pomoni, L. Rastelli, JHEP 1206:072 (2012); P. Liendo, E. Pomoni, L. Rastelli, JHEP, 1209:003 (2012); C. Beem, M. Lemos, P. Liendo, W. Peelaers, L. Rastelli & B.C. van Rees, Commun. Math. Phys. **336** (2015) 1359-1433; C. Beem, L. Rastelli & B.C. van Rees, W Symmetry in six dimensions, arXiv:1404.1079; L. Rastelli, S.S. Razamat, The superconformal index of theories of class S, arXiv:1412.7131; C. Beem, M. Lemos, P. Liendo, L. Rastelli and B.C. van Rees, "The $\mathcal{N} = 2$ superconformal bootstrap," arXiv:1412.7541.

[42] A. Torrielli, J. Geom. Phys. **61**, 230-236 (2011). J. Phys. **A44**, 263001 (2011); M. de Leeuw, T. Matsumoto, S. Moriyama, V. Regelskis & A. Torrielli, Physica Scripta **86**, 028502 (2012)

[43] W.D. Goldberger, W. Skiba & M. Son, Phys. Rev. **D86**, 025019 (2012);
W.D. Goldberger, Z.U. Khandker, Daliang Li & W. Skiba, Phys. Rev. **D88**, 125010 (2013).

[44] T. Andrade and C.F. Uhlemann, JHEP 1201 (2012) 123;
T. Ohl & Ch. F. Uhlemann, JHEP 1205:161 (2012).

[45] T. Creutzig, P. Gao, A. R. Linshaw, JHEP 1204 (2012) 031.

[46] K. Hanaki, C. Peng, JHEP 1308 (2013) 030.

[47] C. -Y. Ju, W. Siegel, Phys. Rev. **D90** (2014) 12, 125004.

[48] A.A. Ardehali, J.T. Liu, P. Szepietowski, JHEP 1306 (2013) 024; JHEP 1402 (2014) 064.

[49] K.H. Neeb, H. Salmasian, Transf. Groups Vol. 18 Issue: 3 (2013) 803.

[50] F. Bonetti, T.W. Grimm and S. Hohenegger, JHEP 1305 (2013) 129,

[51] T. Quella and V. Schomerus, J. Phys. **A46** (2013) 494010, arXiv:1307.7724v2.

[52] M. Buican, JHEP 1401 (2014) 155; M. Buican, T. Nishinaka and C. Papageorgakis, JHEP 2014, 2014:95; M. Buican, T. Nishinaka, arXiv:1410.3006.

[53] V.P. Spiridonov & G.S. Vartanov, Commun. Math. Phys. **325** (2014) 421,

[54] D. Li and A. Stergiou, JHEP, 10 (2014) 037.

[55] M. Beccaria, A.A. Tseytlin, JHEP 1411 (2014) 114.

[56] K. Coulembier, Journal of Algebra 399 (2014) 131-169,

[57] T. Matsumoto and A. Molev, J. Math. Phys. **55** (2014) 091704.

[58] J. Fokken, C. Sieg and M. Wilhelm, JHEP 1407 (2014) 150.

[59] F. Delduc, M. Magro and B. Vicedo, JHEP 1410 (2014) 132.

[60] A. Alldridge, Fréchet globalisations of Harish-Chandra supermodules, arXiv:1403.4055

[61] A. Ghodsi, B. Khavari and A. Naseh, JHEP 1501 (2015) 137

[62] V.G. Kac, Adv. Math. **26**, 8-96 (1977); Comm. Math. Phys. **53**, 31-64 (1977); the second paper is an adaptation for physicists of the first paper.

[63] V.G. Kac, Lect. Notes in Math. **676** (Springer-Verlag, Berlin, 1978) pp. 597-626.

[64] N.N. Shapovalov, Funkts. Anal. Prilozh. **6** (4) 65 (1972); English translation: Funkt. Anal. Appl. **6**, 307 (1972).

[65] I.N. Bernstein, I.M. Gel'fand and S.I. Gel'fand, Funkts. Anal. Prilozh. **5** (1) (1971) 1; English translation: Funct. Anal. Appl. **5** (1971) 1.

[66] V.K. Dobrev, Canonical construction of intertwining differential operators associated with representations of real semisimple Lie groups, Rept. Math. Phys. **25** (1988) 159-181.

[67] V.K. Dobrev, Subsingular vectors and conditionally invariant (q-deformed) equations, J. Phys. A: Math. Gen. **28** (1995) 7135 - 7155.

[68] V.K. Dobrev, Kazhdan-Lusztig polynomials, subsingular vectors, and conditionally invariant (q-deformed) equations, Invited talk at the Symposium "Symmetries in Science IX", Bregenz, Austria, (August 1996), Proceedings, eds. B. Gruber et al, (Plenum Press, New York and London, 1997) pp. 47-80.

[69] V.K. Dobrev, Multiplet classification of the reducible elementary representations of real semi-simple Lie groups: the $SO_e(p,q)$ example, Lett. Math. Phys. **9** (1985) 205-211.

[70] V.K. Dobrev, Lett. Math. Phys. **22** (1991) 251-266.

[71] V.G. Kac, "Characters of typical representations of classical Lie superalgebras", Comm. Algebra **5**, 889-897 (1977).

[72] D. Kazhdan and G. Lusztig, Inv. Math. **53**, 165 (1979).

[73] J. Dixmier, *Enveloping Algebras*, (North Holland, New York, 1977).

[74] V.K. Dobrev and I. Salom, in preparation.

Reflections on Information Paradox in Black Holes

Alexey Golovnev*
Saint Petersburg State University;
Ulyanovskaya ul. 1;
198504 Saint Petersburg, Stary Petergof;
Russia

Abstract

We present some thoughts on the recent developments in the field of information loss problem in Black Hole evaporation. We want to make a point that a possible source of confusion is neglection of the tiniest quantum gravity effects.

Black Holes are unique objects which allow for meaningful theoretical studies of strong gravity and even quantum gravity effects. An infalling and a distant observer would have very different views on the structure of the world. However, a careful analysis has shown that it entails no genuine contradictions for physics, and the paradigm of observer complementarity has been coined. Recently this picture was put into doubt. In particular, it was argued that in old Black Holes a firewall must form in order to protect the basic principles of quantum mechanics. This AMPS paradox has already been discussed in a vast number of papers with different attitudes and conclusions. Here we want to argue that a possible source of confusion is neglection of quantum gravity effects. Contrary to widespread perception, it does not necessarily mean that effective field

* e-mail address: agolovnev@yandex.ru

theory is inapplicable in rather smooth neighbourhoods of large Black Hole horizons. The real offender might be an attempt to consistently use it over the huge distances from the near-horizon zone of old Black Holes to the early radiation. We give simple estimates in support of this viewpoint and show how the Page time and (somewhat more speculative) scrambling time do appear.

It is an amazing fact about Black Holes that they can emit particles [1]. Assuming that this radiation is purely thermal and that a Black Hole will eventually evaporate completely, the information which has been swallowed by the Black Hole during its lifetime is lost, in contradiction to the unitary nature of quantum mechanics. It was then established that unitarity can still be saved if an effective description of the outer near-horizon region is allowed in terms of a Planck-width stretched horizon which can absorb, thermalise and emit information [2]. Of course, due to equivalence principle, an infalling observer should not experience anything special while crossing the horizon. However, there would be no way for him to share this knowledge with the distant colleague, and therefore one can avoid running into a contradiction by adopting the fancy viewpoint that there is no need for a universal global effective field theory description of physics in the whole spacetime. It marked the birth of the Black Hole complementarity paradigm [2].

In order to better understand the relevant time scales of evaporation, recall that a Black Hole of mass M has Hawking temperature

$$\mathrm{k}_B T = \frac{\hbar c^3}{8\pi G M}$$

with the usual notations for the Planck constant $\hbar \equiv \frac{h}{2\pi}$, the speed of light c, the gravitational constant G, and the Boltzmann constant k_B. Under the Stefan-Boltzmann law, the emissive power scales as $T^4 \propto \frac{1}{M^4}$ while the horizon area $\propto M^2$ which implies $\dot{M} \propto \frac{1}{M^2}$ for the rate of energy loss, and therefore the characteristic time of evaporation grows with mass as M^3 [3], or more explicitly $t \sim \left(\frac{M}{m_{Pl}}\right)^3 t_{Pl}$ where the subscript Pl stands for the Planckian quantities.

The problem [4] presents itself when the Black Hole has emitted one half of its total entropy[1], or at the Page time

$$t_{\mathrm{P}} \sim \left(\frac{M}{m_{Pl}}\right)^3 \cdot t_{Pl}.$$

In this case the already emitted radiation contains practically all information which has gone into the Black Hole [8], and therefore must be fully entangled with the late radiation appearing in the near-horizon zone. On the other hand, from the viewpoint of an infalling observer, both sides of the horizon are just the two halves of very flat (almost Minkowski) space. Under the spell of equivalence principle, the freely falling observer has all the legal rights to expect meeting the vacuum state there. And the vacuum state is a very regular thing characterised by maximal entanglement of those two halves. Of course, being simultaneously maximally entangled with two different systems is impossible. The corresponding general principle of quantum mechanics is known under the beautiful name of monogamy of entanglement.

There is apparently a strong tension between the equivalence principle and quantum mechanics which becomes evident at the Page time. But it was also argued [4] that the same should be true much earlier, at the scrambling time when the information of constituent matter has already been scrambled (thermalised) by the Black Hole.

[1]Note that, under the name of energy curtains, this paradox was earlier reported in Ref. [5], see also Refs. [6, 7].

This time scale is much smaller. Actually, there are some reasonably good reasons to believe that it must be of order

$$t_{scr} \sim \left(\frac{M}{m_{Pl}} \log \frac{M}{m_{Pl}}\right) \cdot t_{Pl}$$

which makes the Black Holes amazingly fast scramblers [9]. The status of this time scale is however unclear [10].

With one time scale or another, the way out proposed in Ref. [4] was a firewall just behind the horizon. In other words, the equivalence principle is sacrificed in such a way that we can not trust the usual effective field theory in the zone. Adopting entanglement with the early radiation, we can no longer afford entanglement between the two halves of the near-horizon region which translates into presence of energetic quanta around the horizon, or a firewall.

A natural attempt to fully identify the interior with the distant radiation is not only extremely non-local but also leads to the frozen vacuum [11] or inability of the infalling observer to excite the near-horizon state which violates the equivalence principle no less than a firewall.

A possible alternative would be to resort to strong complementarity by arguing that it is not a problem when two observers see absolutely different physical content of the zone if they cannot communicate their findings to each other. It does not seem to work out in an ideal way because an infalling observer might perform a precise measurement of the early radiation before entering the zone, or because there is some time for a freely falling observer to change his mind and turn around inside the zone before crossing the horizon [12]. However, the contradictory measurements and inferences might turn to be computationally unfeasible [13], extremely fine-grained [14], overwhelmingly affecting the Black Hole state, or even be akin to observing quantum superpositions of macroscopic worlds [15].

The problem remains controversial, and it is only clear

that new insights are needed and, hopefully, expected. They would certainly deepen our understanding of quantum mechanics and gravity; and the relevant issues are really exciting. For example, one line of reasoning [16] asserts that the very measurement made by the distant observer creates high energy quanta, or a firewall, which would kill the infalling colleague. If this process is to be causally conceivable, and if we do insist on causality of so violent behaviours, then Einstein-Rosen bridges between the Black Hole and its early radiation must be invoked [17].

Instead, we would like to offer a different approach to the AMPS problem which might actually point at better integrity of physical description. We argue that the paradoxes might be resolved by taking quantum gravity effects into account in the form of unavoidable entanglement with microscopic geometrical configurations of spacetime.

Our main idea is that locally an effective field theory description can be valid and very precise everywhere in the low curvature regions, although taken all the way over huge spacetime distances, the tiny errors might accumulate considerably enough to entail the loss of purity of the early Hawking radiation. Note, for an illustration, that mean calendar year length difference of Gregorian and Julian calendars is less than eleven minutes and at first glance seems impractical, but in four hundred years it sums up to three full days. Random errors do not grow as fast as a gradual change but still can eventually matter.

Below we give some simple estimates of quantum gravity effects on propagating radiation, and show how the Page and scrambling times can naturally appear from such considerations.

Let us first address the clear-cut problem at the Page time $t_{\mathrm{P}} \propto M^3$. If we want to take the quantum gravity effects into account, then probably it would be safest and fairly model-independent to assume that the wavelength of a typical photon of Hawking radiation

$$\lambda \sim \frac{M}{m_{Pl}} \cdot l_{Pl}$$

cannot be determined with precision better than the Planck length l_{Pl}. We treat it as an intrinsic fluctuation of the wavelength

$$\delta\lambda \sim l_{Pl}.$$

If the photon has propagated over a huge number N of wavelengths, then the statistical uncertainty of the path length $L = N\lambda$ amounts to

$$\delta L \sim \sqrt{N} \cdot l_{Pl}$$

from which we see that δL reaches λ when

$$L \sim \frac{\lambda^3}{l_{Pl}^2} \sim \left(\frac{M}{m_{Pl}}\right)^3 \cdot l_{Pl} \sim ct_{\mathrm{P}}.$$

Therefore, to the Page time, the information about the relative phases of different photons is definitely lost. Of course, the reason is that we treat the geometry as a mute background arena for electrodynamics. But the actual quantum state of photons gets dynamically entangled with quantum fluctuations of geometry and, after a long enough time, tracing over the states of geometry produces a very blurred image of the emitted light. Since it is no longer pure, there are no obstructions for the late radiation to be entangled with the Black Hole interiors. Quantum gravity has gone into full play despite the incredible smallness of all its local effects.

The scrambling time $t_{scr} \propto M \log M$ is much trickier to discuss in this context. But we argue that a more careful treatment of the decoherence features allows to naturally come at this time scale, too. Let us assume that the dynamics of photons in presence of small quantum gravitational corrections can be described as an open quantum system with the Lindblad equation [18, 19]

$$\dot{\rho} = \hat{\mathcal{L}}\rho$$

for the density matrix ρ. It provides a natural framework for discussing phenomenology of quantum gravity, see for example the paper [20] and references therein for possible effects in the oscillations of neutral kaons. The dynamical semigroup generator $\hat{\mathcal{L}}$ consists of two terms: the commutator with the Hamiltonian which reproduces the standard Schrödinger equation and the additional Lindblad operator, the required properties of which we do not need to discuss now.

The order of magnitude of the coefficients in the matrix of the Lindblad operator depends on the adopted level of coarse graining. We want to address situations in which an effective field theory description is just marginally enough to come to a contradiction. It is natural then to adopt a resolution at the level of ultraviolet cutoff scale which is presumably the Planck scale. According to our considerations above, we can expect to start confusing the neighbouring states after the time period $\tau \sim \frac{\lambda}{c}$, and therefore the matrix entries of the Lindblad operator are expected of order $\frac{c}{\lambda} \sim \frac{m_{Pl}}{Mt_{Pl}}$.

Note then that we are not interested in simply converting one state into another with off-diagonal elements of the Lindblad operator because it does not automatically entail decoherence, even though it may produce very interesting effects such as CPT-violation [20]. We would rather like to find an independent growth of probabilities for other states bringing the system to a statistical mixture. For the initial Cauchy data, we can assume that a given photon has been in a given pure state with the probability p_1 practically equal to 1. However, due to quantum gravity effects, the other states could not have been totally absent. Their initial probabilities $p_j(0)$ can be estimated as $\sim \frac{m_{Pl}}{M}$, or some mild power of it. We see that typically $\log p_j$ starts growing with time as $\frac{m_{Pl}}{Mt_{Pl}} \cdot t$. And extrapolating this trend far beyond any reasonable limit, an undoubtedly mixed state is reached when

$$\log \frac{M}{m_{Pl}} \sim \frac{m_{Pl}}{Mt_{Pl}} \cdot t,$$

or at the fast scrambling time. Admittedly, it sounds rather speculative, but so is the issue of scrambling time itself in the context of firewalls.

Of course, it was always clear that somehow relaxing the assumption of entanglement between the early and the late radiation would give a way to resolving the paradox [21]. Moreover, a concrete realisation was proposed in the context of Many Worlds interpretation of quantum mechanics [22]. A Black Hole randomly emits really huge amounts of low-energy quanta, and therefore, for an old specie, its position must be very indefinite due to recoil effects [3]. Accordingly, we have to face macroscopic superpositions in the system. (We note in passing, it might be interesting to compare these superpositions with those which appear in the Ref. [15].) It is certainly a logical possibility that after specifying a certain branch for the macroscopic world the unitarity is lost despite being safe in the full picture [23]. However, it seems rather radical an idea which probably could make macroscopic quantum superpositions (too easily) observable.

Our proposal is different. We talk about small fluctuations of geometry irrespective to the foundational issues of quantum mechanics. Note also that what we mean is not just gravitons radiated from the Black Hole which are of little or no interest for resolving the paradox, but it is really an effect of a slightly random medium with the spacetime foam on the way of the photons. Of course, it would be completely legitimate to wonder how the distant fluctuations and the nearby emission effects contrive to avoid the potential tensions and save the day.

Unfortunately, we are not so much aware of the details of quantum gravity and even its real degree of non-locality. However, our point is that these non-local quantum gravity effects need not be locally observable with any deviations from effective field theory or with other conceivable types of anomalies such as the frozen states. What matters is only the Planck scale physics.

Although we are very far from giving the final and definitive solutions, the estimates look very interesting because they produce the relevant time scales from a completely different side. Actually, it is not the first time when Black Holes teach us non-trivial lessons about the Nature. Very remarkably, Black Holes obey the usual laws of thermodynamics [24] which should not be expected of a simple and fairly isolated system. Somehow, general relativity has given the hints to a deeper parent theory which has to describe Black Holes as statistical systems with many degrees of freedom. And now we might learn some new lessons about quantum gravity regimes from an unexpected direction. We ought to be ready and open-minded for new insights. Black Holes have a good credit history, and it would be a good idea to take seriously what they say.

Acknowledgements. The author is grateful to A. Andrianov for useful discussions. The author was partially supported by Saint Petersburg State University research grant No. 11.38.660.2013 and by Russian Foundation for Basic Research grant No. 12-02-31214.

References

[1] S.W. Hawking, *Comm. Math. Phys.* **43** (1975), 199.

[2] L. Susskind, L. Thorlacius, J. Uglum, *Phys. Rev. D* **48** (1993), 3743; *arXiv:* hep-th/9306069.

[3] D.N. Page, *Phys. Rev. Lett.* **44** (1980), 301.

[4] A. Almheiri, D. Marolf, J. Polchinski, J. Sully, *J. High Energy Phys.* JHEP02(2013)062; *arXiv:* 1207.3123.

[5] S.L. Braunstein, *arXiv:* 0907.1190v1.

[6] S.L. Braunstein, S. Pirandola, K. Zyczkowski, *Phys. Rev. Lett.* **110** (2013), 101301; *arXiv:* 0907.1190v3.

[7] S.L. Braunstein, S. Pirandola, *arXiv:* 1311.1326.

[8] D.N. Page, *Phys. Rev. Lett.* **71** (1993), 1291; *arXiv:* gr-qc/9305007.

[9] Y. Sekino, L. Susskind, *J. High Energy Phys.* JHEP10(2008)065; *arXiv:* 0808.2096.

[10] L. Susskind, *arXiv:* 1208.3445.

[11] R. Bousso, *Phys. Rev. Lett.* **112** (2014), 041102; *arXiv:* 1308.3697.

[12] R. Bousso, *Phys. Rev. D* **87** (2013), 124023; *arXiv:* 1207.5192.

[13] D. Harlow, P. Hayden, *J. High Energy Phys.* JHEP06(2013)085; *arXiv:* 1301.4504.

[14] L. Susskind, *arXiv:* 1311.7379.

[15] Y. Nomura, J. Varela, S.J. Weinberg, *J. High Energy Phys.* JHEP03(2013)059; *arXiv:* 1207.6626.

[16] L. Susskind, *arXiv:* 1311.3335.

[17] J. Maldacena, L. Susskind, *Fortschritte der Physik* **61** (2013), 781; *arXiv:* 1306.0533.

[18] G. Lindblad, *Comm. Math. Phys.* **48** (1976), 119.

[19] V. Gorini, A. Kossakowski, E.C.G. Sudarshan, *J. Math. Phys.* **17** (1976), 821.

[20] A.A. Andrianov, R. Tarrach, J. Taron, *Phys. Lett. B* **507** (2001), 200; *arXiv:* hep-ph/0010276.

[21] S. Hossenfelder, *arXiv:* 1210.5317.

[22] S.D.H. Hsu, *arXiv:* 1302.0451.

[23] S.D.H. Hsu, *arXiv:* 1308.5686.

[24] J.D. Bekenstein, *Phys. Rev. D* **9** (1974), 3292.

Unification of Inflation and Dark Energy from Spontaneous Breaking of Scale Invariance

Eduardo Guendelman*
Physics Department, Ben Gurion University of the Negev
Beer Sheva, Israel

Emil Nissimov and Svetlana Pacheva†
Institute for Nuclear Research and Nuclear Energy
Bulgarian Academy of Sciences, Sofia, Bulgaria

ABSTRACT

We propose a new class of gravity-matter models defined in terms of two independent non-Riemannian volume forms (alternative generally covariant integration measure densities) on the spacetime manifold. For the matter we choose appropriate scalar field potentials of exponential form so that the full gravity-matter system is invariant under global Weyl-scale symmetry. Solution of the pertinent equations of motion produce two dimensionful integration constants which spontaneously break global Weyl-scale invariance. In the resulting effective Einstein-frame gravity-matter system we obtain an effective potential for the scalar matter field which has an interesting cosmological application, namely, it allows for a unified description of both an early universe inflation and present day dark energy.

1. Introduction

A component of the energy-momentum tensor of matter which is proportional to the spacetime metric tensor, with the proportionality constant being indeed exactly or approximately a spacetime constant, has been widely discussed and its consequences understood, but the possible origin of such terms remains a subject of hot discussion.

One can transfer such energy-momentum tensor component to the left hand side of Einstein's equations and then it can be considered as belonging to the gravity part. This was the way indeed how Einstein introduced such contribution and named it the "cosmological constant term".

More recently it has been invoked as a fundamental component of the energy density of both the early universe and of the present universe. Nowadays we call such component "vacuum energy density". The vacuum energy density has been used as the source of a possible inflationary phase of the early universe (the pioneering papers on the subject are [1]; for a non-technical review and a good collection of further references on different

* e-mail address: guendel@bgu.ac.il

† e-mail address: nissimov@inrne.bas.bg, svetlana@inrne.bas.bg

aspects of inflation see Ref.[2]; for a more technical review see Ref.[3]). Inflation provides an attractive scenario for solving some of the fundamental puzzles of the standard Big Bang model, like the horizon and the flatness problems (third ref.[1]) as well as providing a framework for sensible calculations of primordial density perturbations (for a review, see the book [4]).

Also, with the discovery of the accelerating expansion of the present universe (for reviews of this subject, see for example [5, 6]) it appears plausible that a small vacuum energy density, usually referred in this case as "dark energy", is also present even today. Because of this discovery the cosmological constant problem (CCP) has evolved from the "Old Cosmological Constant Problem" [7], where physicists were concerned with explaining why the observed vacuum energy density of the universe nowadays is vanishing, to a different type of CCP – the "New Cosmological Constant Problem" [8]. Namely, the problem now is to explain why the vacuum energy density of the current universe is very small rather than being zero.

These two vacuum energy densities, the one of inflation and the other of the universe nowadays, have however a totally different scale. One then wonders how cosmological evolution may naturally interpolate between such two apparently quite distinctive physical situations.

The possibility of continuously connecting an inflationary phase to a slowly accelerating universe through the evolution of a single scalar field – the *quintessential inflation scenario* – has been first studied in Ref.[9]. Also, carefully constructed F(R) models can yield both an early time inflationary epoch and a late time de Sitter phase with vastly different values of effective vacuum energies [10]. For a recent proposal of a quintessential inflation mechanism based on "variable gravity" model [11] and for extensive list of references to earlier work on quintessential inflation, see Ref.[12].

In the present letter we propose a new theoretical framework where the quintessential inflation scenario is explicitly realized in a natural way.

The main idea of our current approach comes from Refs.[13, 14, 15] (for recent developments, see Refs.[16]), where some of us have proposed a new class of gravity-matter theories based on the idea that the action integral may contain a new metric-independent integration measure density, *i.e.*, an alternative non-Riemannian volume form on the spacetime manifold defined in terms of an auxiliary antisymmetric gauge field of maximal rank. The latter formalism yields various new interesting results in all types of known generally coordinate-invariant theories:

- (i) $D = 4$-dimensional models of gravity and matter fields containing the new measure of integration appear to be promising candidates for resolution of the dark energy and dark matter problems, the fifth force problem, and a natural mechanism for spontaneous breakdown of global Weyl-scale symmetry [13]-[16].
- (ii) Study of reparametrization invariant theories of extended objects (strings and branes) based on employing of a modified non-Riemannian world-sheet/world-volume integration measure [17] leads to dynamically induced variable string/brane tension and to string models of non-abelian confinement.

- (iii) Study in Refs.[18] of modified supergravity models with an alternative non-Riemannian volume form on the spacetime manifold produces some outstanding new features: (a) This new formalism applied to minimal $N = 1$ supergravity naturally triggers the appearance of a dynamically generated cosmological constant as an arbitrary integration constant, which signifies a new explicit mechanism of spontaneous (dynamical) breaking of supersymmetry; (b) Applying the same formalism to anti-de Sitter supergravity allows us to appropriately choose the above mentioned arbitrary integration constant so as to obtain simultaneously a very small effective observable cosmological constant as well as a very large physical gravitino mass.

We now extend the above formalism employing two (instead of only one) modified non-Riemannian volume-forms on the underlying spacetime to construct new type of gravity-matter models producing interesting cosmological implications relating inflationary and slowly accelerating phases of the universe.

2. Gravity-Matter Models With Two Independent Non-Riemannian Volume-Forms

We shall consider the following non-standard gravity-matter system with an action of the general form (for simplicity we will use units where the Newton constant is taken as $G_N = 1/16\pi$):

$$S = \int d^4x\,\Phi_1(A)\Big[R + L^{(1)}\Big] + \int d^4x\,\Phi_2(B)\Big[L^{(2)} + \frac{\Phi(H)}{\sqrt{-g}}\Big]\,, \tag{1}$$

with the following notations:

- $\Phi_1(A)$ and $\Phi_2(B)$ are two independent non-Riemannian volume-forms, *i.e.*, generally covariant integration measure densities on the underlying spacetime manifold:

$$\Phi_1(A) = \frac{1}{3!}\varepsilon^{\mu\nu\kappa\lambda}\partial_\mu A_{\nu\kappa\lambda} \quad , \quad \Phi_2(B) = \frac{1}{3!}\varepsilon^{\mu\nu\kappa\lambda}\partial_\mu B_{\nu\kappa\lambda}\,, \tag{2}$$

  defined in terms of field-strengths of two auxiliary 3-index antisymmetric tensor gauge fields. $\Phi_{1,2}$ take over the role of the standard Riemannian integration measure density $\sqrt{-g} \equiv \sqrt{-\det\|g_{\mu\nu}\|}$ in terms of the spacetime metric $g_{\mu\nu}$.

- $R = g^{\mu\nu}R_{\mu\nu}(\Gamma)$ and $R_{\mu\nu}(\Gamma)$ are the scalar curvature and the Ricci tensor in the first-order (Palatini) formalism, where the affine connection $\Gamma^\mu_{\nu\lambda}$ is *a priori* independent of the metric $g_{\mu\nu}$.

- $L^{(1,2)}$ denote two different Lagrangians with matter fields, to be specified below.

- $\Phi(H)$ indicate the dual field strength of a third auxiliary 3-index antisymmetric tensor gauge field:

$$\Phi(H) = \frac{1}{3!}\varepsilon^{\mu\nu\kappa\lambda}\partial_\mu H_{\nu\kappa\lambda}; , \tag{3}$$

whose presence is crucial for non-triviality of the model.

For the matter Lagrangians we take the scalar field ones:

$$L^{(1)} = -\frac{1}{2}g^{\mu\nu}\partial_\mu\varphi\partial_\nu\varphi - V(\varphi) \quad , \quad L^{(2)} = U(\varphi) \ \text{(no kinetic term)} \, . \tag{4}$$

We now observe that the original action (1) is invariant under global Weyl-scale transformations:

$$g_{\mu\nu} \to \lambda g_{\mu\nu} \ , \ \varphi \to \varphi - \frac{1}{\alpha}\ln\lambda \ ,$$
$$A_{\mu\nu\kappa} \to \lambda A_{\mu\nu\kappa} \ , \ B_{\mu\nu\kappa} \to \lambda^2 B_{\mu\nu\kappa} \ , \ H_{\mu\nu\kappa} \to H_{\mu\nu\kappa} \ , \tag{5}$$

where α is a dimensionful positive parameter, provided we choose the scalar field potentials in (4) in the form (similar to the choice [13]):

$$V(\varphi) = f_1 \exp\{-\alpha\varphi\} \quad , \quad U(\varphi) = f_2 \exp\{-2\alpha\varphi\} \, . \tag{6}$$

Variation of (1) w.r.t. $\Gamma^\mu_{\nu\lambda}$ gives (following the derivation in [13]):

$$\Gamma^\mu_{\nu\lambda} = \Gamma^\mu_{\nu\lambda}(\bar{g}) = \frac{1}{2}\bar{g}^{\mu\kappa}\left(\partial_\nu\bar{g}_{\lambda\kappa} + \partial_\lambda\bar{g}_{\nu\kappa} - \partial_\kappa\bar{g}_{\nu\lambda}\right) \, , \tag{7}$$

where $\bar{g}_{\mu\nu}$ is the Weyl-rescaled metric:

$$\bar{g}_{\mu\nu} = \chi_1 g_{\mu\nu} \ , \ \chi_1 \equiv \frac{\Phi_1(A)}{\sqrt{-g}} \, . \tag{8}$$

Variation of the action (1) w.r.t. auxiliary tensor gauge fields $A_{\mu\nu\lambda}$, $B_{\mu\nu\lambda}$ and $H_{\mu\nu\lambda}$ yields the equations:

$$\partial_\mu\Big[R + L^{(1)}\Big] = 0 \quad , \quad \partial_\mu\Big[L^{(2)} + \frac{\Phi(H)}{\sqrt{-g}}\Big] = 0 \quad , \quad \partial_\mu\Big(\frac{\Phi_2(B)}{\sqrt{-g}}\Big) = 0 \, , \tag{9}$$

whose solutions read:

$$\frac{\Phi_2(B)}{\sqrt{-g}} = \chi_2 = \text{const} \quad , \quad R + L^{(1)} = -M_1 = \text{const} \ ,$$
$$L^{(2)} + \frac{\Phi(H)}{\sqrt{-g}} = -M_2 = \text{const} \, . \tag{10}$$

Here M_1 and M_2 are arbitrary dimensionful and χ_2 arbitrary dimensionless integration constants. The appearance of M_1, M_2 signifies *dynamical spontaneous breakdown* of global Weyl-scale invariance under (5) due to the scale non-invariant solutions (second and third ones) in (10).

Varying (1) w.r.t. $g_{\mu\nu}$ and using relations (10) we have:

$$\chi_1 \Big[R_{\mu\nu} + \frac{\partial}{\partial g^{\mu\nu}} L^{(1)} \Big] - \frac{1}{2} \chi_2 \Big[T^{(2)}_{\mu\nu} + g_{\mu\nu} M_2 \Big] = 0 \,, \tag{11}$$

where χ_1 and χ_2 are defined in (8) and first relation (10), and $T^{(2)}_{\mu\nu}$ is the energy-momentum tensor of the second matter Lagrangian with the standard definitions:

$$T^{(1,2)}_{\mu\nu} = g_{\mu\nu} L^{(1,2)} - 2 \frac{\partial}{\partial g^{\mu\nu}} L^{(1,2)} \,. \tag{12}$$

Using second relation (10) and (12), Eqs.(11) can be put in the form:

$$R_{\mu\nu} - \frac{1}{2} g_{\mu\nu} R = \frac{1}{2} \Big[T^{(1)}_{\mu\nu} + g_{\mu\nu} M_1 + \frac{\chi_2}{\chi_1} \Big(T^{(2)}_{\mu\nu} + g_{\mu\nu} M_2 \Big) \Big] \,. \tag{13}$$

Taking the trace of Eqs.(13) and using again second relation (10) we solve for the scale factor χ_1:

$$\chi_1 = 2 \chi_2 \frac{U(\varphi) + M_2}{V(\varphi) - M_1} \,. \tag{14}$$

Now, taking into account (8) and (14) we can bring Eqs.(13) into the standard form of Einstein equations for the metric $\bar{g}_{\mu\nu}$ (8) , *i.e.*, the Einstein frame equations:

$$R_{\mu\nu}(\bar{g}) - \frac{1}{2} \bar{g}_{\mu\nu} R(\bar{g}) = \frac{1}{2} T^{\rm eff}_{\mu\nu} \tag{15}$$

with energy-momentum tensor corresponding (according to (12)) to the following effective scalar field Lagrangian:

$$L_{\rm eff} = - \frac{1}{2} \bar{g}^{\mu\nu} \partial_\mu \varphi \partial_\nu \varphi - U_{\rm eff}(\varphi) \,, \tag{16}$$

where the effective scalar field potential reads:

$$U_{\rm eff}(\varphi) = \frac{\left(V(\varphi) - M_1\right)^2}{4\chi_2 (U(\varphi) + M_2)} = \frac{\left(f_1 e^{-\alpha\varphi} - M_1\right)^2}{4\chi_2 \left(f_2 e^{-2\alpha\varphi} + M_2\right)} \,. \tag{17}$$

3. Canonical Hamiltonian Treatment

Before proceeding to the cosmological implications of the new gravity-matter model based on two non-Riemannian spacetime volume forms let us briefly discuss the application of the canonical Hamiltonian formalism to (1), which will elucidate the proper physical meaning of the arbitrary integration constants χ_2, M_1, M_2 (10) encountered in the previous section.

For convenience let us introduce the following short-hand notations for the field-strengths (2), (3) of the auxiliary 3-index antisymmetric gauge fields $A_{\mu\nu\lambda}$, $B_{\mu\nu\lambda}$, $H_{\mu\nu\lambda}$ (the dot indicating time-derivative):

$$\Phi_1(A) = \dot{A} + \partial_i A^i \quad , \quad A = \frac{1}{3!}\varepsilon^{ijk} A_{ijk} \quad , \quad A^i = -\frac{1}{2}\varepsilon^{ijk} A_{0jk} \, , \tag{18}$$

$$\Phi_1(B) = \dot{B} + \partial_i B^i \quad , \quad B = \frac{1}{3!}\varepsilon^{ijk} B_{ijk} \quad , \quad B^i = -\frac{1}{2}\varepsilon^{ijk} B_{0jk} \, , \tag{19}$$

$$\Phi(H) = \dot{H} + \partial_i H^i \quad , \quad H = \frac{1}{3!}\varepsilon^{ijk} H_{ijk} \quad , \quad H^i = -\frac{1}{2}\varepsilon^{ijk} H_{0jk} \, , \tag{20}$$

For the pertinent canonical momenta we have:

$$\pi_A = R + L^{(1)} \equiv \widetilde{L}_1(u, \dot{u}) \quad , \quad \pi_B = L^{(2)}(u, \dot{u}) + \frac{1}{\sqrt{-g}}(\dot{H} + \partial_i H^i) \, ,$$
$$\pi_H = \frac{1}{\sqrt{-g}}(\dot{B} + \partial_i B^i) \, , \tag{21}$$

where $(u, \dot{u})$ collectively denote the set of the basic gravity-matter canonical variables $((u) = (g_{\mu\nu}, \varphi, \ldots))$ and their velocities, and:

$$\pi_{A^i} = 0 \quad , \quad \pi_{B^i} = 0 \quad , \quad \pi_{H^i} = 0 \, , \tag{22}$$

the latter implying that A^i, B^i, H^i will in fact appear as Lagrange multipliers for certain first-class Hamiltonian constraints (see Eqs.(26)-(27) below). For the canonical momenta conjugated to the basic gravity-matter canonical variables we have (using last relation (21)):

$$p_u = (\dot{A} + \partial_i A^i)\frac{\partial}{\partial \dot{u}}\widetilde{L}_1(u, \dot{u}) + \pi_H \sqrt{-g}\frac{\partial}{\partial \dot{u}} L^{(2)}(u, \dot{u}) \, . \tag{23}$$

Now, from relations (21), (23) we obtain the velocities $\dot{u}, \dot{A}, \dot{B}, \dot{H}$ as functions of canonically conjugate momenta $\dot{u}{=}\dot{u}\,(u, \pi_A, \pi_B, \pi_H)$ etc. Taking into account (21)-(22) (and the short-hand notations (18)-(20)) the canonical Hamiltonian corresponding to (1):

$$\mathcal{H} = p_u\, \dot{u} + \pi_A\, \dot{A} + \pi_B\, \dot{B} + \pi_H\, \dot{H} - (\dot{A} + \partial_i A^i)\widetilde{L}_1(u, \dot{u})$$
$$- \pi_H \sqrt{-g}\Big[L^{(2)}(u, \dot{u}) + \frac{1}{\sqrt{-g}}(\dot{H} + \partial_i H^i)\Big] \tag{24}$$

acquires the following form as function of the canonically conjugated variables (here $\dot{u}=\dot{u}\,(u,\pi_A,\pi_B,\pi_H)$):

$$\mathcal{H} = p_u\,\dot{u} - \pi_H\sqrt{-g}L^{(2)}(u,\dot{u}) + \sqrt{-g}\pi_H\pi_B \\ -\partial_i A^i\pi_A - \partial_i B^i\pi_B - \partial_i H^i\pi_H \; . \tag{25}$$

We are interested only in the canonical Hamiltonian structure related to the auxiliary antisymmetric tensor gauge fields. From the second line in (25) we deduce that indeed A^i, B^i, H^i are Lagrange multipliers for the first-class Hamiltonian constraints:

$$\partial_i\pi_A = 0 \;\; \to \;\; \pi_A = -M_1 = \text{const} \; , \tag{26}$$

and similarly:

$$\pi_B = -M_2 = \text{const} \quad , \quad \pi_H = \chi_2 = \text{const} \; , \tag{27}$$

which are the canonical Hamiltonian counterparts of Lagrangian constraint equations of motion (10).

Thus, the canonical Hamiltonian treatment of (1) reveals the meaning of the auxiliary 3-index antisymmetric tensor gauge fields $A_{\mu\nu\lambda}$, $B_{\mu\nu\lambda}$, $H_{\mu\nu\lambda}$ – building blocks of the non-Riemannian spacetime volume-form formulation of the modified gravity-matter model (1). Namely, the canonical momenta π_A, π_B, π_H conjugated to the "magnetic" parts A, B, H (18)-(20) of the auxiliary 3-index antisymmetric tensor gauge fields are constrained through Dirac first-class constraints (26)-(27) to be constants identified with the arbitrary integration constants χ_2, M_1, M_2 (10) arising within the Lagrangian formulation of the model. The canonical momenta π^i_A, π^i_B, π^i_H conjugated to the "electric" parts A^i, B^i, H^i (18)-(20) of the auxiliary 3-index antisymmetric tensor gauge field are vanishing (22) which makes the latter canonical Lagrange multipliers for the above Dirac first-class constraints.

4. Implications for Cosmology

The effective scalar potential (17) possesses the following remarkable property. For large negative and large positive values of φ $U_{\text{eff}}(\varphi)$ exponentially fast approaches two infinitely large flat regions (which we will denote as $(\mp)$ flat regions, respectively) with smooth transition between them:

$$U_{\text{eff}}(\varphi) \to \frac{f_1^2}{4\chi_2 f_2} \quad \text{for} \;\; \varphi \to -\infty \; , \\ U_{\text{eff}}(\varphi) \to \frac{M_1^2}{4\chi_2 M_2} \quad \text{for} \;\; \varphi \to +\infty \; . \tag{28}$$

The shape of $U_{\text{eff}}(\varphi)$ depicted on Fig.1.

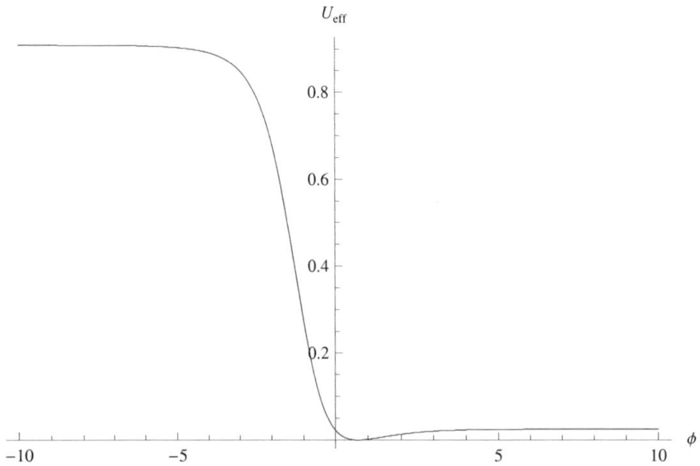

Figure 1: Shape of the effective scalar potential $U_{\text{eff}}(\varphi)$ (17).

In the original gravity-matter models with only one non-Riemannian volume form [13] one obtains upon spontaneous breakdown of global Weyl-scale symmetry only one flat region of the effective scalar potential, so that this simple model does not meet the requirement for unification of inflation and dark energy.

Let us point out that in the context of the original modified-measure gravity-matter theories (with only one non-Riemannian integration measure density) it is possible to obtain two flat regions by means of adding an ϵR^2 term as shown in [15]. This is, however, achieved at the price of creating a non-canonical kinetic term for the scalar field which substantially complicates the theory and its particle content interpretation (see remarks on this point below).

In the present case we derived an effective scalar potential $U_{\text{eff}}(\varphi)$ (17) with *two infinitely large flat regions* while the kinetic term of the scalar field remained canonical. In the course of the derivation we obtained three integration constants χ_2, M_1, M_2 (10), two of them (M_1, M_2) triggering spontaneous breakdown of the original global Weyl-scale symmetry (5). These integration constants can be appropriately adjusted so as to get the shape of the effective scalar potential as depicted on Fig.1.

The cosmological picture suggested by Fig.1 is evident. The universe starts from a large negative value of φ, then slow rolls the $(-)$ flat region to the left whose height:

$$U_{\text{eff}}(\varphi) \simeq U_{(-)} = \frac{f_1^2}{4\chi_2 f_2} \tag{29}$$

upon appropriate choice of f_1, f_2 can be made very large corresponding to the vacuum energy density in the inflationary phase. After that there is an abrupt fall to $U_{\rm eff} = 0$ where particle creation is obtained from rapidly varying $\varphi(t)$. The scalar field comes down with very high kinetic energy in the region of $U_{\rm eff} \simeq 0$, certainly higher than the value of $U_{\rm eff}$ in the (+) flat region to the right:

$$U_{\rm eff}(\varphi) \simeq U_{(+)} = \frac{M_1^2}{4\chi_2 M_2} , \tag{30}$$

which upon appropriate choice of the scales of M_1, M_2 (see below) can be made to correspond to the correct value of the current vacuum energy density. So $\varphi(t)$ "climbs" the latter low barrier and continues to evolve in the $\varphi \to +\infty$ direction. Thus, on the (+) flat region we have a slow rolling scalar which produces approximately the dark energy equation of state ($\rho \simeq -p$, with very small $\rho \simeq U_{(+)} = \frac{M_1^2}{4\chi_2 M_2}$) explaining the present day dark energy phase.

Indeed, taking the integration constant $\chi_2 \sim 1$, and choosing the scales of the scalar potential (17) coupling constants $M_1 \sim M_{EW}^4$ and $M_2 \sim M_{Pl}^4$, where M_{EW}, M_{Pl} are the electroweak and Plank scales, respectively, we are then naturally led to a very small vacuum energy density of the order:

$$\rho \simeq U_{(+)} \sim M_{EW}^8/M_{Pl}^4 \sim 10^{-120} M_{Pl}^4 , \tag{31}$$

which gives the right order of magnitude for the present epoche's vacuum energy density as already recognized in Ref.[19].

In a parallel work [20] we have generalized the model (1) by including a gravitational R^2 term so as to preserve the original global Weyl-scale symmetry (5):

$$S = \int d^4x\, \Phi_1(A)\Big[R + L^{(1)}\Big] + \int d^4x\, \Phi_2(B)\Big[L^{(2)} + \epsilon R^2 + \frac{\Phi(H)}{\sqrt{-g}}\Big] . \tag{32}$$

The analysis of the model (32) goes along similar lines as described in Sections 2 and 3 above, where in addition we find for a definite parameter range a non-singular "emergent universe" solution which describes an initial phase of universe's evolution that precedes the inflationary phase. It was also realized in [20] that upon taking the order of magnitude for the coupling constants in the effective scalar potential $f_1 \sim f_2 \sim (10^{-2}M_{Pl})^4$, then the order of magnitude of the vacuum energy density of the early universe $U_{(-)}$ (29) becomes:

$$U_{(-)} \sim f_1^2/f_2 \sim 10^{-8} M_{Pl}^4 , \tag{33}$$

which conforms to the BICEP2 experiment [21] and Planck Collaboration data [22] implying the energy scale of inflation of order $10^{-2}M_{Pl}$. Nevertheless, as shown in [20], the result for the tensor-to-scalar ratio r obtained within the model (32) conforms to the data of the Planck Collaboration [22] rather than BICEP2 [21].

Acknowledgments.

E.G. and E.N. are sincerely grateful to Prof. Branko Dragovich and the organizers of the *Eight Meeting in Modern Mathematical Physics* in Belgrade for cordial hospitality. We are thankful to Alexander Kaganovich for stimulating discussions, as well as to Ramon Herrera and Pedro Labrana for a fruitful collaboration on the extension of the current project. We gratefully acknowledge support of our collaboration through the academic exchange agreement between the Ben-Gurion University and the Bulgarian Academy of Sciences. E.N. and S.P. are partially supported by Bulgarian NSF Grant *DFNI T02/6*. S.P. has received partial support from European COST action MP-1210.

References

[1] A. Starobinsky, *Phys. Lett.* **91B** (1980) 99-102;
K. Sato, *Mon. Not. Roy. Astron. Soc.* **195** (1981) 467-479;
A.H. Guth, *Phys. Rev.* **D23** (1981) 347;
A.D. Linde, *Phys. Lett.* **108B** (1982) 389;
A. Albrecht and P.J. Steinhardt, *Phys. Rev. Lett.* **48** (1982) 1220.

[2] A. Guth, *"The Inflationary Universe"*, Vintage, Random House (1998).

[3] E.W. Kolb and M.S. Turner, *"The Early Universe"*, Addison Wesley (1990).

[4] S. Dodelson, *"Modern Cosmology"*, Acad. Press (2003).

[5] M.S. Turner, in *Third Stromle Symposium "The Galactic Halo"*, ASP Conference Series Vol.**666**, B.K. Gibson, T.S. Axelrod and M.E. Putman (eds.), 1999;
N. Bahcall, J.P. Ostriker, S.J. Perlmutter and P.J. Steinhardt, *Science* **284** (1999) 1481;
for a review, see P.J.E. Peebles and B. Ratra, *Rev. Mod. Phys.* **75** (2003) 559.

[6] A. Riess, *et al.*, *Astronomical Journal* 116 (1998) 1009-1038;
S. Perlmutter *et al.*, *Astrophysical Journal* 517 (1999) 565-586.

[7] S. Weinberg, *The Cosmological Constant Problem*, *Rev. Mod. Phys.* **61** (1989) 1;
Y.Jack Ng, *The Cosmological Constant Problem*, *Int. J. Mod. Phys.* **D1** (1992) 145.

[8] S. Weinberg, *The Cosmological constant problems*, in Proceedings of Conference *Dark Matter 2000*, p.18-26 (*arxiv:astro-ph/0005265*);
see also S. Weinberg, *"Cosmology"*, Oxford Univ. Press (2008).

[9] P.J.E. Peebles and A.Vilenkin, *Phys. Rev.* **D59** (1999) 063505.

[10] S.A. Appleby, R.A. Battye and A.A. Starobinsky, *JCAP* 1006 (2010) 005 (*arXiv:0909.1737* [astro-ph]).

[11] C. Wetterich, *Phys. Rev.* **D89** (2014) 024005 (*arxiv:1308.1019* [astro-ph]).

[12] Md. Wali Hossain, R. Myrzakulov, M. Sami and E.N. Saridakis, *Phys. Rev.* **D90** (2014) 023512 (*arxiv:1402.6661* [gr-qc]).

[13] E.I. Guendelman, *Mod. Phys. Lett.* **A14** (1999) 1043-1052 (*arxiv:gr-qc/9901017*).

[14] E.I. Guendelman and A. Kaganovich, *Phys. Rev.* **D60** (1999) 065004 (*arxiv:gr-qc/9905029*).

[15] E.I. Guendelman and O. Katz, *Class. Quantum Grav.* **20** (2003) 1715-1728 (*arxiv:gr-qc/0211095*).

[16] E.I. Guendelman and P. Labrana, *Int. J. Mod. Phys.* **D22** (2013) 1330018 (*arxiv:1303.7267* [astro-ph.CO]);
E.I. Guendelman, D. Singleton and N. Yongram, *JCAP* 1211 (2012) 044 (*arxiv:1205.1056* [gr-qc]);
E.I. Guendelman, H.Nishino and S. Rajpoot, *Phys. Lett.* **732B** (2014) 156 (*arXiv:1403.4199* [hep-th]).

[17] E. Guendelman, A. Kaganovich, E. Nissimov and S. Pacheva, *Phys. Rev.* **D66** (2002) 046003 (*arxiv:hep-th/0203024*).

[18] E. Guendelman, E.Nissimov, S. Pacheva and M. Vasihoun, *Bulg. J. Phys.* **40** (2013) 121-126 (*arxiv:1310.2772* [hep-th]);
E. Guendelman, E.Nissimov, S. Pacheva and M. Vasihoun, *arxiv:1404.4733* [hep-th].

[19] N. Arkani-Hamed, L.J. Hall, C. Kolda and H. Murayama, *Phys. Rev. Lett.* **85** (2000) 4434-37 (*astro-ph/0005111*).

[20] E. Guendelman, R. Herrera, P. Labrana, E. Nissimov and S. Pacheva, *Emergent Cosmology, Inflation and Dark Energy*, *Gen. Rel. Grav.* **47** (2015) # 2, DOI: 10.1007/s10714-015-1852-1 (*arxiv:1408.5344v4* [hep-th], 2015)

[21] P.A.R. Ade *et.al.* (BICEP2 Collaboration), *Phys. Rev. Lett.* **112** (2014) 241101 (*arxiv:1403.3985*) [astro-ph.CO]).

[22] R. Adam *et al.* (Planck Collaboration), *arxiv:1409.5738* [astro-ph.CO];
R. Adam *et al.* (Planck Collaboration), *arxiv:1303.5082* [astro-ph.CO].

A New Venue of Spontaneous Supersymmetry Breaking in Supergravity

Eduardo Guendelman and Mahary Vasihoun *
Physics Department, Ben Gurion University of the Negev
Beer Sheva, Israel

Emil Nissimov and Svetlana Pacheva†
Institute for Nuclear Research and Nuclear Energy
Bulgarian Academy of Sciences, Sofia, Bulgaria

Abstract

We present a qualitatively new mechanism for dynamical spontaneous breakdown of supersymmetry in supergravity. Specifically, we construct a modified formulation of standard minimal $N = 1$ supergravity as well as of anti-de Sitter supergravity in terms of a non-Riemannian spacetime volume form (generally covariant integration measure density). The new supergravity formalism naturally triggers the appearance of a *dynamically generated cosmological constant* as an arbitrary integration constant which signifies *spontaneous (dynamical) breaking of supersymmetry.* Applying the new formalism to anti-de Sitter supergravity allows us to appropriately choose the above mentioned arbitrary integration constant so as to obtain simultaneously a *very small effective observable cosmological constant* as well as a *large physical gravitino mass* as required by modern cosmological scenarios for slowly expanding universe of the present epoch.

1. Introduction

Supersymmetry is a fundamental extended space-time symmetry of Nature, believed to manifest itself at ultra-high energies, which is unifying bosons (integer-spin particles) and fermions (half-integer spin particles). Among the prinicipal theoretical highlights of supersymmetry one should mention: drastically reducing the number of apriori independent physical parameters; drastically reducing (in some cases even eliminating) ultraviolet divergencies in quantum field theories; providing possible solution of the hierachy/fine-tuning problems in high-energy elementary particles phenomenology; naturally appearing within the context of modern string theory. For a very short list of basic references see [1, 2, 3, 4]. Apart from elementary particle physics and cosmology, supersymmetry plays an increasingly active role both as conceptional paradigm and as important

* e-mail address: guendel@bgu.ac.il, mahary@bgu.ac.il
† e-mail address: nissimov@inrne.bas.bg, svetlana@inrne.bas.bg

mathematical tool in various other principal areas of theoretical physics and mathematics such as theoretical condensed matter [5] as well as in the modern theory of integrable (soliton) systems [6].

Unfortunately, supersymmetry is *not an exact* symmetry of Nature. Otherwise, we would observe bosonic (integer-spin) counterparts with the same masses of the fundamental fermionic particles – protons, electrons, *etc.*. Therefore, supersymmetry must be *spontaneously broken* [1, 2].

Under spontaneous symmetry breakdown the symmetry-generating charges in the respective (quantum) field theories are conserved whereas the ground states ("vacuums") are *not* invariant under the symmetry transformations. Spontaneous symmetry breakdown is always accompanied by the appearance of certain mass (energy) scale of the breakdown. Typically, the scale of spontaneous symmetry breaking is generated by the appearance of non-zero vacuum expectation values of certain (quantum) fields non-trivially transforming under the pertinent symmetry group. The basic example is the *Brout-Englert-Higgs* mechanism in the Standard Model of particle physics (see *e.g.* Ref.[7]).

In minimal supergravity (supersymmetric generalizations of ordinary Einstein general relativity without interactions with other matter fields – for a recent account of modern supergravity theories and notations, see Ref.[4]) there is another way to spontaneously break supersymmetry (supersymmetric Brout-Englert-Higgs effect) – via *dynamical generation of non-zero cosmological constant* [8, 9]. In what follows we describe a new theoretical framework where the above scenario is explicitly realized in a natural way.

The main idea of our current approach comes from Refs.[10] (for recent developments, see Refs.[11]), where some of us have proposed a new class of gravity-matter theories based on the idea that the action integral may contain several terms built with *different* spacetime volume-forms (generally covariant integration measure densities). Namely, apart from terms built with the standard Riemannian volume-form given in terms of the square-root of the determinant of the pertinent Riemannian metric, we may have additional term(s) constructed with one or more [13] *alternative non-Riemannian volume form(s) on the spacetime manifold* in terms of one or more auxiliary antisymmetric tensor gauge field(s) of maximal rank completely independent of the metric. The latter formalism has lead to various new interesting results in all types of known generally coordinate-invariant theories:

- (i) $D = 4$-dimensional models of gravity and matter fields containing one term with an alternative non-Riemannian integration measure appear to be promising candidates for resolution of the dark energy and dark matter problems, the fifth force problem, and a natural mechanism for spontaneous breakdown of global Weyl-scale symmetry [10]-[11].
- (ii) Study of reparametrization invariant theories of extended objects (strings and branes) based on employing of a modified non-Riemannian world-sheet/world-volume integration measure [12] leads to dynamically induced variable string/brane tension and to simple string models

of non-abelian confinement.

- (iii) In Refs.[13] a new class of extended gravity-matter models is constructed, built in terms of two independent non-Riemannian volume-forms on the underlying spacetime manifold, producing interesting cosmological implications relating inflationary and today's slowly accelerating phases of the universe.

The principal results described in the next sections are as follows:

(a) First, we briefly outline the main properties of a general class of gravity-matter models with one non-Riemannian volume-form term. Specifically we provide a consistent canonical Hamiltonian analysis of the latter models exhibiting the physical meaning of the pertinent auxiliary fields which are absent in the standard formulation of gravity-matter actions in terms of the ordinary Riemannian spacetime volume-form (see also second Ref.[13]).

(b) The new non-Riemannian volume-form formalism is then applied to minimal $N = 1$ supergravity in $D = 4$-dimensional spacetime. This naturally triggers the appearance of a *dynamically generated* cosmological constant as an arbitrary integration constant, which signifies a new explicit mechanism of spontaneous (dynamical) breaking of supersymmetry.

(c) Applying the same formalism to *anti-de Sitter* supergravity allows us to appropriately choose the above mentioned arbitrary integration constant so as to obtain simultaneously a very small effective observable cosmological constant as well as a very large physical gravitino mass.

2. Gravity-Matter Models With a Non-Riemannian Spacetime Volume-Form

Let us consider the following non-standard gravity-matter system of a general form whose action is a linear combination (with $c_{1,2}$ – some constants) of one term with an alternative non-Riemannian spacetime volume-form and another one with the standard Riemannian one:

$$S = c_1 \int d^D x\, \Phi(B) \Big[L^{(1)} + \Phi(H) \Big] + c_2 \int d^D x \sqrt{-g}\, L^{(2)} \tag{1}$$

Here the following notations are used:

- The alternative volume element in the first term of (1) is given by the following *non-Riemannian* integration measure density:

$$\Phi(B) \equiv \frac{1}{(D-1)!} \varepsilon^{\mu_1 \ldots \mu_D}\, \partial_{\mu_1} B_{\mu_2 \ldots \mu_D} \, , \tag{2}$$

where $B_{\mu_1 \ldots \mu_{D-1}}$ is an auxiliary rank $(D-1)$ antisymmetric tensor gauge field. The latter can also be parametrized in terms of D auxiliary scalar fields $B_{\mu_1 \ldots \mu_{D-1}} = \frac{1}{D} \varepsilon_{I J_1 \ldots J_{D-1}} \phi^I \partial_{\mu_1} \phi^{J_1} \ldots \partial_{\mu_{D-1}} \phi^{J_{D-1}}$, so that: $\Phi(B) = \frac{1}{D!} \varepsilon^{\mu_1 \ldots \mu_D}\, \varepsilon_{I_1 \ldots I_D} \partial_{\mu_1} \phi^{I_1} \ldots \partial_{\mu_D} \phi^{I_D}$, but we will stick to the definition (2). The volume element in the second term of (1) is given by the standard Riemannian integration measure density $\sqrt{-g}$, where $g \equiv \det \|g_{\mu\nu}\|$ is the determinant of the corresponding Riemannian metric $g_{\mu\nu}$.

- The Lagrangians $L^{(1,2)} \equiv \frac{1}{2\kappa^2} R + L^{(1,2)}_{\text{matter}}$ include both standard Einstein-Hilbert gravity action as well as matter/gauge-field parts. Here $R = g^{\mu\nu} R_{\mu\nu}(\Gamma)$ is the scalar curvature within the first-order (Palatini) formalism and $R_{\mu\nu}(\Gamma)$ is the Ricci tensor in terms of the independent affine connection $\Gamma^{\mu}_{\lambda\nu}$.
- In general, the second Lagrangian $L^{(2)}$ might contain also higher curvature terms like R^2 (see e.g. first Ref.[13]).
- In the first *modified-measure term* of the action (1) we have included an additional term containing the field-strength $\Phi(H)$ of another auxiliary rank $(D-1)$ antisymmetric tensor gauge field $H_{\mu_1 \ldots \mu_{D-1}}$:

$$\Phi(H) \equiv \frac{1}{(D-1)!} \varepsilon^{\mu_1 \ldots \mu_D} \, \partial_{\mu_1} H_{\mu_2 \ldots \mu_D} \,, \tag{3}$$

whose presence is crucial for non-triviality of the model. Such term would be purely topological (total divergence) one if included in standard Riemannian integration measure action like the second term with $L^{(2)}$ on the r.h.s. of (1).

The auxiliary gauge fields $B_{\mu_1 \ldots \mu_{D-1}}$ and $H_{\mu_1 \ldots \mu_{D-1}}$ turn out to be pure-gauge (non-propagating) degrees of freedom, however, both are leaving remnants which play crucial role in the sequel (see also next Section). Namely, varying (1) w.r.t. H and B tensor gauge fields we get:

$$\partial_\mu \Big(\frac{\Phi(B)}{\sqrt{-g}}\Big) = 0 \;\; \to \;\; \frac{\Phi(B)}{\sqrt{-g}} \equiv \chi = \text{const} \,, \tag{4}$$

$$\partial_\mu \Big[L^{(1)} + \Phi(H)\Big] = 0 \;\; \to \;\; L^{(1)} + \Phi(H) = M = \text{const} \,, \tag{5}$$

where χ (ratio of the two measure densities) and M are *arbitrary integration constants*. The meaning of χ and M from the point of view of canonical Hamiltonian formalism is elucidated in the next section.

Now, varying (1) w.r.t. $g^{\mu\nu}$ and taking into account (4)–(5) we arrive at the following effective Einstein equations (in the first-order formalism):

$$R_{\mu\nu}(\Gamma) - \frac{1}{2} g_{\mu\nu} R + \Lambda_{\text{eff}} g_{\mu\nu} = \kappa^2 T^{\text{eff}}_{\mu\nu}, \tag{6}$$

with effective energy-momentum tensor:

$$T^{\text{eff}}_{\mu\nu} = g_{\mu\nu} L^{\text{eff}}_{\text{matter}} - 2 \frac{\partial L^{\text{eff}}_{\text{matter}}}{\partial g^{\mu\nu}} \;\; , \;\; L^{\text{eff}}_{\text{matter}} \equiv \frac{1}{c_1 \chi + c_2} \Big[c_1 L^{(1)}_{\text{matter}} + c_2 L^{(2)}_{\text{matter}}\Big] \,, \tag{7}$$

and with a *dynamically generated effective cosmological constant* $\Lambda_{\rm eff}$ thanks to the non-zero integration constants M, χ:

$$\Lambda_{\rm eff} = \kappa^2 \left(c_1\chi + c_2\right)^{-1} \chi M \,. \tag{8}$$

3. Canonical Hamiltonian Treatment

In what follows we restrict our attention to $D = 4$-dimensional spacetime. For convenience we will introduce the following short-hand notations for the field-strengths (2) and (3) of the auxiliary 3-index antisymmetric gauge fields $B_{\mu\nu\lambda}$, $H_{\mu\nu\lambda}$ (the dot indicating time-derivative):

$$\Phi(B) = \dot{B} + \partial_i B^i \quad , \quad B = \frac{1}{3!}\varepsilon^{ijk} B_{ijk} \;, \quad B^i = -\frac{1}{2}\varepsilon^{ijk} B_{0jk} \;, \tag{9}$$

$$\Phi(H) = \dot{H} + \partial_i H^i \quad , \quad H = \frac{1}{3!}\varepsilon^{ijk} H_{ijk} \;, \quad H^i = -\frac{1}{2}\varepsilon^{ijk} H_{0jk} \;, \tag{10}$$

According to the general form of the action (1) (for simplicity we set here $c_{1,2} = 1$) the pertinent canonically conjugated momenta read:

$$\pi_B = L^{(1)}(u, \dot{u}) + \frac{1}{\sqrt{-g}}(\dot{H} + \partial_i H^i) \,, \qquad \pi_H = \frac{1}{\sqrt{-g}}(\dot{B} + \partial_i B^i) \,, \tag{11}$$

where $(u, \dot{u})$ collectively denote the set of the basic gravity-matter canonical variables $((u) = (g_{\mu\nu}$, matter fields$))$ and their velocities, and:

$$\pi_{B^i} = 0 \quad , \quad \pi_{H^i} = 0 \,. \tag{12}$$

Eqs.(12) imply that B^i, H^i will in fact appear as Lagrange multipliers for certain first-class Hamiltonian constraints (see Eqs.(16) below).

Using (11), for the canonical momenta conjugated to the basic gravity-matter canonical variables we obtain:

$$p_u = \pi_H \frac{\partial}{\partial \dot{u}}\Bigl(\sqrt{-g}L^{(1)}(u, \dot{u})\Bigr) + \frac{\partial}{\partial \dot{u}}\Bigl(\sqrt{-g}L^{(2)}(u, \dot{u})\Bigr) \,. \tag{13}$$

Now, from (11) and (13) we obtain the velocities $\dot{H} = \dot{H}\,(u, p_u, \pi_H, \pi_B)$, $\dot{B} = \dot{B}\,(u, \pi_H)$ and $\dot{u} = \dot{u}\,(u, p_u, \pi_H, \pi_B)$ as functions of the respective canonically conjugate momenta, wherefrom the canonical Hamiltonian corresponding to (1):

$$\mathcal{H} = p_u\, \dot{u} + \pi_B\, \dot{B} + \pi_H\, \dot{H} \\ -(\dot{B} + \partial_i B^i)\Bigl[L^{(1)}(u, \dot{u}) + \frac{1}{\sqrt{-g}}(\dot{H} + \partial_i H^i)\Bigr] - \sqrt{-g}L^{(2)}(u, \dot{u}) \tag{14}$$

acquires the following form as function of the canonically conjugated variables (here $\dot{u}$=$\dot{u}$ $(u, p_u, \pi_H, , \pi_B)$:

$$\mathcal{H} = p_u \; \dot{u} \; - \pi_H \sqrt{-g} L^{(1)}(u, \dot{u}) - \sqrt{-g} L^{(2)}(u, \dot{u}) \\ + \sqrt{-g} \pi_H \pi_B - \partial_i B^i \pi_B - \partial_i H^i \pi_H \; . \tag{15}$$

From (15) we deduce that indeed B^i, H^i are Lagrange multipliers for the first-class Hamiltonian constraints:

$$\pi_H = \chi = \mathrm{const} \quad , \quad \pi_B = M = \mathrm{const} \; , \tag{16}$$

which (in virtue of (11)) are the canonical Hamiltonian counterparts of Lagrangian constraint equations of motion (4)-(5).

We conclude that the canonical Hamiltonian treatment of (1) reveals the meaning of the auxiliary 3-index antisymmetric tensor gauge fields $B_{\mu\nu\lambda}$, $H_{\mu\nu\lambda}$. Namely, the canonical momenta π_B, π_H conjugated to the "magnetic" parts B, H (9)-(10) of the respective tensor gauge fields are constrained through Dirac first-class constraints (16) to be constants identified with the arbitrary integration constants χ, M (4)-(5) arising within the Lagrangian formulation of the model. The canonical momenta π^i_B, π^i_H conjugated to the "electric" parts B^i, H^i (9)-(10) of the auxiliary 3-index antisymmetric tensor gauge field are vanishing (12) which makes the latter canonical Lagrange multipliers for the above Dirac first-class constraints.

4. Supersymmetric Brout-Englert-Higgs Effect in Minimal Supergravity

Let us now apply the above formalism to construct a non-Riemannian spacetime volume-form version of simplest $N = 1$ supergravity in $D = 4$.

Let us recall the standard component-field action of $D = 4$ minimal $N = 1$ supergravity (for definitions and notations we follow [4]):

$$S_{\mathrm{SG}} = \frac{1}{2\kappa^2} \int d^4x \, e \Big[R(\omega, e) - \bar{\psi}_\mu \gamma^{\mu\nu\lambda} D_\nu \psi_\lambda \Big] \; , \tag{17}$$

$$e = \det \| e^a_\mu \| \quad , \quad R(\omega, e) = e^{a\mu} e^{b\nu} R_{ab\mu\nu}(\omega) \; . \tag{18}$$

$$R_{ab\mu\nu}(\omega) = \partial_\mu \omega_{\nu ab} - \partial_\nu \omega_{\mu ab} + \omega^c_{\mu a} \omega_{\nu cb} - \omega^c_{\nu a} \omega_{\mu cb} \; . \tag{19}$$

$$D_\nu \psi_\lambda = \partial_\nu \psi_\lambda + \frac{1}{4} \omega_{\nu ab} \gamma^{ab} \psi_\lambda \quad , \quad \gamma^{\mu\nu\lambda} = e^\mu_a e^\nu_b e^\lambda_c \gamma^{abc} \; , \tag{20}$$

where all objects belong to the first-order "vierbein" (frame-bundle) formalism.

The vierbeins e^a_μ (describing the graviton) and the spin-connection $\omega_{\mu ab}$ ($SO(1,3)$ gauge field acting on the gravitino ψ_μ) are *a priori* independent fields (their relation arises subsequently on-shell); $\gamma^{ab} \equiv \frac{1}{2} \left(\gamma^a \gamma^b - \gamma^b \gamma^a \right)$ *etc.* denote antisymmetrized products of gamma-matrices with γ^a being the ordinary Dirac gamma-matrices. The invariance of the action (17) under local supersymmetry transformations:

$$\delta_\epsilon e^a_\mu = \frac{1}{2}\bar{\varepsilon}\gamma^a\psi_\mu \quad , \quad \delta_\epsilon\psi_\mu = D_\mu\varepsilon \tag{21}$$

follows from the invariance of the pertinent Lagrangian density up to a total derivative:

$$\delta_\epsilon\Big(e[R(\omega,e) - \bar{\psi}_\mu\gamma^{\mu\nu\lambda}D_\nu\psi_\lambda]\Big) = \partial_\mu[e(\bar{\varepsilon}\zeta^\mu)] \,, \tag{22}$$

where ζ^μ functionally depends on the gravitino field ψ_μ.

We now propose a modification of (17) by replacing the standard generally-covariant measure density $e = \sqrt{-g}$ by the alternative measure density $\Phi(B)$ (Eq.(2) for $D = 4$):

$$\Phi(B) \equiv \frac{1}{3!}\varepsilon^{\mu\nu\kappa\lambda}\,\partial_\mu B_{\nu\kappa\lambda} \,, \tag{23}$$

and we will use the general framework described above. The modified supergravity action reads:

$$S_{\rm mSG} = \frac{1}{2\kappa^2}\int d^4x\,\Phi(B)\left[R(\omega,e) - \bar{\psi}_\mu\gamma^{\mu\nu\lambda}D_\nu\psi_\lambda + \frac{\varepsilon^{\mu\nu\kappa\lambda}}{3!\,e}\,\partial_\mu H_{\nu\kappa\lambda}\right] \,, \tag{24}$$

where a new term containing the field-strength (Eq.(3) for $D = 4$) of a 3-index antisymmetric tensor gauge field $H_{\nu\kappa\lambda}$ has been added.

The equations of motion w.r.t. $H_{\nu\kappa\lambda}$ and $B_{\nu\kappa\lambda}$ yield:

$$\partial_\mu\Big(\frac{\Phi(B)}{e}\Big) = 0 \;\to\; \frac{\Phi(B)}{e} \equiv \chi = \text{const} \,, \tag{25}$$

$$R(\omega,e) - \bar{\psi}_\mu\gamma^{\mu\nu\lambda}D_\nu\psi_\lambda + \frac{\varepsilon^{\mu\nu\kappa\lambda}}{3!\,e}\,\partial_\mu H_{\nu\kappa\lambda} = 2M \,, \tag{26}$$

where χ and M are arbitrary integration constants.

The action (24) is invariant under local supersymmetry transformations (21) supplemented by transformation laws for $H_{\mu\nu\lambda}$ and $\Phi(B)$:

$$\delta_\epsilon H_{\mu\nu\lambda} = -e\,\varepsilon_{\mu\nu\lambda\kappa}(\bar{\varepsilon}\zeta^\kappa) \quad , \quad \delta_\epsilon\Phi(B) = \frac{\Phi(B)}{e}\,\delta_\epsilon e \,, \tag{27}$$

which algebraically close on-shell, *i.e.*, when Eq.(25) is imposed.

The appearance of the integration constant M represents a *dynamically generated cosmological constant* in the pertinent gravitational equations of motion and, thus, it signifies a *spontaneous (dynamical) breaking of supersymmetry.*

Indeed, varying (24) w.r.t. e^a_μ:

$$e^{b\nu} R^a_{b\mu\nu} - \frac{1}{2}\bar{\psi}_\mu \gamma^{a\nu\lambda} D_\nu \psi_\lambda + \frac{1}{2}\bar{\psi}_\nu \gamma^{a\nu\lambda} D_\mu \psi_\lambda$$

$$+\frac{1}{2}\bar{\psi}_\lambda \gamma^{a\nu\lambda} D_\nu \psi_\mu + \frac{e^a_\mu}{2} \frac{\varepsilon^{\rho\nu\kappa\lambda}}{3!\, e} \partial_\rho H_{\nu\kappa\lambda} = 0 \qquad (28)$$

and using Eq.(26) (containing the arbitrary integration constant M) to replace the last H-term on the l.h.s. of (28), the results is as follows: we obtain the vierbein counterparts of the Einstein equations including a dynamically generated *floating* cosmological constant term $e^a_\mu M$:

$$e^{b\nu} R^a_{b\mu\nu} - \frac{1}{2} e^a_\mu R(\omega, e) + e^a_\mu M = \kappa^2 T^a_\mu \ , \qquad (29)$$

$$\kappa^2 T^a_\mu \equiv \frac{1}{2}\bar{\psi}_\mu \gamma^{a\nu\lambda} D_\nu \psi_\lambda - \frac{1}{2} e^a_\mu \bar{\psi}_\rho \gamma^{\rho\nu\lambda} D_\nu \psi_\lambda - \frac{1}{2}\bar{\psi}_\nu \gamma^{a\nu\lambda} D_\mu \psi_\lambda - \frac{1}{2}\bar{\psi}_\lambda \gamma^{a\nu\lambda} D_\nu \psi_\mu \ .$$

Let us recall at this point that according to the classic paper [9] the sole appearance of a cosmological constant in supergravity, even in the absence of a manifest mass term for the gravitino, implies that the gravitino becomes *massive*, i.e., it absorbs the Goldstone fermion of spontaneous supersymmetry breakdown – a *supersymmetric Brout-englert-Higgs effect.*

A significantly more interesting scenario occurs when applying the above formalism with non-Riemannian spacetime volume-forms to anti-de Sitter (AdS) supergravity. Namely, let us start with the standard AdS supergravity action (see *e.g.* Ref.[4]):

$$S_{\rm AdS-SG} = \frac{1}{2\kappa^2} \int d^4 x\, e \Big[R(\omega, e) - \bar{\psi}_\mu \gamma^{\mu\nu\lambda} D_\nu \psi_\lambda - m\, \bar{\psi}_\mu \gamma^{\mu\nu} \psi_\nu - 2\Lambda_0 \Big] \quad (30)$$

$$m \equiv \frac{1}{L} \quad , \quad \Lambda_0 \equiv -\frac{3}{L^2} \ . \qquad (31)$$

The action (30) contains additional explicit mass term for the gravitino as well as a bare cosmological constant Λ_0 balanced in a precise way $|\Lambda_0| = 3m^2$ so as to maintain local supersymmetry invariance and, in particular, keeping the *physical gravitino mass zero* in spite of the presence of a "bare" gravitino mass term!

At this point let us stress that here we have AdS spacetime as a background ("vacuum") with curvature radius L (unlike Minkowski background in the absence of a bare cosmological constant). Therefore, the notions of "mass" and "spin" are now given in terms of the eigenvalues of the Casimirs of the unitary irreducible representations (discrete series) of the group of motion of AdS space $SO(2,3) \sim Sp(4,\mathbb{R})$ (for $D = 4$) instead of the ordinary Poincare group ($SO(1,3) \ltimes R^4$) Casimirs (see *e.g.* Ref.[14]). Thus, on AdS background it is possible for the gravitino to be massive even in the absence of a "bare" gravitino mass term and, *vice-versa*, it will be massless even in the presence of a large "bare" mass provided it is tuned up w.r.t. AdS cosmological constant as in (31).

Now, following the same steps as with (24) we construct a modified AdS supergravity with non-Riemannian spacetime volume element:

$$S_{\text{mod-AdS-SG}} = \frac{1}{2\kappa^2}\int d^4x\, \Phi(B)\Big[R(\omega,e) - \bar{\psi}_\mu \gamma^{\mu\nu\lambda} D_\nu \psi_\lambda \\ -m\,\bar{\psi}_\mu \gamma^{\mu\nu}\psi_\nu - 2\Lambda_0 + \frac{\varepsilon^{\mu\nu\kappa\lambda}}{3!\,e}\,\partial_\mu H_{\nu\kappa\lambda}\Big]\,, \qquad (32)$$

with $\Phi(B)$ as in (23) and m, Λ_0 as in (30). The action (32) is invariant under local supersymmetry transformations:

$$\delta_\epsilon e^a_\mu = \frac{1}{2}\bar{\varepsilon}\gamma^a \psi_\mu \quad , \quad \delta_\epsilon \psi_\mu = \Big(D_\mu - \frac{1}{2L}\gamma_\mu\Big)\varepsilon\,, \\ \delta_\epsilon H_{\mu\nu\lambda} = -e\,\varepsilon_{\mu\nu\lambda\kappa}(\bar{\varepsilon}\zeta^\kappa) \quad , \quad \delta_\epsilon \Phi(B) = \frac{\Phi(B)}{e}\,\delta_\epsilon e\,. \qquad (33)$$

The modified AdS supergravity action (32) will trigger dynamical spontaneous supersymmetry breaking resulting in the appearance of the dynamically generated floating cosmological constant M as in Eq.(29) which will add to the bare cosmological constant Λ_0. Now we can use the freedom in choosing the value of the *a priori* arbitrary integration constant M in order to match two basic requirements by modern cosmological scenarios for slowly expanding universe of today [15]. Namely, we can achieve via appropriate choice of $M \simeq |\Lambda_0| = 3m^2$ a *very small effective observable cosmological constant*:

$$\Lambda_{\text{eff}} = M + \Lambda_0 = M - 3m^2 \ll |\Lambda_0| \qquad (34)$$

and, simultaneously, a *large physical gravitino mass* m_{eff}:

$$m_{\text{eff}} \simeq m = \sqrt{\frac{1}{3}|\Lambda_0|}\,, \qquad (35)$$

which will be very close to the large "bare" gravitino mass parameter $m = \sqrt{|\Lambda_0|/3}$ since now because of the smallness of Λ_{eff} (34) the background spacetime geometry becomes almost flat.

5. Conclusions

We have shown that applying the formalism with non-Riemannian spacetime volume-forms in gravity/matter theories provides a simple mechanism for dynamical generation of a cosmological constant. The latter appears as a conserved Dirac-constrained canonical momentum conjugated to the auxiliary maximal-rank antisymmetric gauge field building up the non-Riemannian volume-form. In the context of modified minimal $N = 1$ supergravity defined in terms of a non-Riemannian spacetime volume-form the dynamically generated cosmological constant triggers spontaneous supersymmetry breakdown and gravitino mass generation (supersymmetric

Brout-Englert-Higgs effect). Upon constructing modified anti-de Sitter supergravity with a non-Riemannian spacetime volume-form we can fine-tune the dynamically generated cosmological integration constant in order to achieve simultaneously a very small physical observable cosmological constant and a very large physical observable gravitino mass – a paradigm of modern cosmological scenarios for slowly expanding universe of the present epoch.

Acknowledgments.

E.G. and E.N. are sincerely thankful to Prof. Branko Dragovich and the organizers of the *Eight Meeting in Modern Mathematical Physics* in Belgrade for cordial hospitality. We gratefully acknowledge support of our collaboration through the academic exchange agreement between the Ben-Gurion University and the Bulgarian Academy of Sciences. E.N. and S.P. are partially supported by Bulgarian NSF Grant *DFNI T02/6*. S.P. has received partial support from European COST action MP-1210.

References

[1] J. Wess and J. Bagger, *"Supersymmetry and Supergravity"*, 2ed., Princeton Univ. Press (1992);
S.J. Gates, M.T. Grisaru, M. Rocek and W. Siegel, *"Superspace"*, Addison Wesley (1983);
A.S. Galperin, E.A. Ivanov, V.I. Ogievetsky and E.S. Sokatchev, *"Harmonic Superspace"*, Cambridge Univ. Press (2004);
S. Duplij, W. Siegel and J. Bagger (eds.), *"Concise Encyclopedia of Supersymmetry and Noncommutative Structures in Mathematics and Physics"*, Kluwer Acad. Publ. (2004).

[2] S. Weinberg, *"Quantum Theory of Fields"* vol. 3. "Supersymmetry", Cambridge Univ. Press (2000).

[3] M.B. Green, J.H. Schwarz and E. Witten, *"Superstring Theory"*. vol. 1,2 , Cambridge Univ. Press (1987);
K. Becker, M. Becker and J.H. Schwarz, *"String Theory and M-Theory"*, Cambridge Univ. Press (2007).

[4] D. Freedman and A. van Proeyen, *"Supergravity"*, Cambridge Univ. Press (2012).

[5] K. Efetov, *"Supersymmetry in Disorder and Chaos"*, Cambridge Univ. Press (1997).

[6] H. Aratyn, T. Jimbo. W.-Y. Keung and U. Sukhatme (eds.), *"Supersymmetry and Integrable Models"*, Springer, N.Y. (1998).

[7] S. Weinberg, *"Quantum Theory of Fields"* vol. 1 "Foundations", Cambridge Univ. Press (1995);
S. Weinberg, *"Quantum Theory of Fields"* vol. 2 "Modern Applications", Cambridge Univ. Press (1996).

[8] D.V. Volkov and V.A. Soroka, *J.E.T.P. Letters* **18** (1973) 312;
D.V. Volkov and V.A. Soroka, *Theor. Math. Phys.* **20** (1974) 829.

[9] S. Deser and B. Zumino, *Phys. Rev. Lett.* **38** (1977) 1433-1436.

[10] E.I. Guendelman, *Mod. Phys. Lett.* **A14** (1999) 1043-1052 (arxiv:*gr-qc/9901017*);

E.I. Guendelman and A. Kaganovich, *Phys. Rev.* **D60** (1999) 065004 (*arxiv:gr-qc/9905029*);
E.I. Guendelman and O. Katz, *Class. Quantum Grav.* **20** (2003) 1715-1728 (*arxiv:gr-qc/0211095*).

[11] E.I. Guendelman and P. Labrana, *Int. J. Mod. Phys.* **D22** (2013) 1330018 (*arxiv:1303.7267* [astro-ph.CO]);
E.I. Guendelman, D. Singleton and N. Yongram, *JCAP* 1211 (2012) 044 (*arxiv:1205.1056* [gr-qc]);
E.I. Guendelman, H.Nishino and S. Rajpoot, *Phys. Lett.* **732B** (2014) 156 (*arXiv:1403.4199* [hep-th]).

[12] E. Guendelman, A. Kaganovich, E. Nissimov and S. Pacheva, *Phys. Rev.* **D66** (2002) 046003 (*arxiv:hep-th/0203024*).

[13] E. Guendelman, R. Herrera, P. Labrana, E. Nissimov and S. Pacheva, *Gen. Rel. Grav.* **47** (2015) #2, DOI: 10.1007/s10714-015-1852-1 (*arxiv:1408.5344v4* [hep-th], 2015);
E. Guendelman, E. Nissimov and S. Pacheva, *Unification of Inflation and Dark Energy from Spontaneous Breaking of Scale Invariance*, this Proceedings (*arxiv:1407.6281v3* [hep-th], 2015).

[14] J.P. Gazeau and M. Novello, *J. Phys. A: Math. Theor.* **41** (2008) 304008.

[15] Adam G. Riess *et.al.* (Supernova Search Team), *Astronomical Journal* **116** (1998) 1009-1038 (*arxiv:astro-ph/9805201*);
S. Perlmutter *et.al.* (The Supernova Cosmology Project), *Astrophysical Journal* **517** (1999) 565-586 (*arxiv:astro-ph/9812133*);
A.G. Riess *et.al.*, *Astrophysical Journal* **607** (2004) 665-687 (*arxiv:astro-ph/0402512*).

Another modification of the ACD sum rule

S. R. Ignjatović*
Department of Physics, Prirodno-matematički fakultet,
Mladena Stojanovića 2, 78000 Banja Luka, Bosnia and Herzegovina

ABSTRACT

The Analytic Continuation by Duality (ACD) method is used to estimate dispersive integrals in low-energy QCD phenomenology and in technicolor models. The method uses a polynomial approximation of the $1/x$ function, usually a best-L_p approximation. There are several sources of error in the ACD estimates which, along with the oscillatory behavior of the best-Lp approximations, render them unreliable. The method is unstable, but it is still occasionally used in QCD phenomenology. Here we investigate a modification of the ACD method that has recently appeared by using a simple model spectrum. The modified ACD uses an approximation weighted towards the end of the interval, where the subasymptotic QCD is expected to be more reliable. For the case when only the first few terms of the Operator Product Expansion (OPE) are known, the modified method fails to reproduce the results expected for the simple model.

1. Introduction

The Analytic Continuation by Duality (ACD) is a method used to evaluate dispersive integrals in QCD using the first few terms of the Operator Product Expansion (OPE). ACD differs considerably from the usual QCD sum rules, but nevertheless it is often termed a "finite energy sum rule". ACD is a fairly old method [2], used mainly in low-energy QCD phenomenology.

The ACD approximates the kernel of a dispersive integral by a polynomial on the real line, which allows the integration to be performed in the complex plane. As noted early [3], ACD is an ill-posed problem: small variations in the input parameters may lead to large variations in the results, which is discussed in one of our previous contributions [5]. Besides, we have found the ACD to be unstable and unreliable [4, 8]. The instability seems to be related to the oscillatory character of the best-L_p polynomial approximations. It is difficult to investigate the reliability of the ACD method, thus, we have to be content with an analysis of its behavior on simple model spectra for which the exact value of the esimate is found from a dispersive integral and compared to the ACD-derived estimate.

Several modifications of the ACD method have been proposed. In one of the modifications [6] – called 'simplified' ACD here – the $1/s$ kernel is approximated by truncated Taylor series around the upper cut-off R. That

* e-mail address: sinisha@teol.net

simplified method is stable, but (contrary to claims) it yields only a lower bound for the estimates [7]. We have attempted a version of the ACD with an asymmetric weighting function used for the polynomial approximation, but it proved to be less reliable than the original ACD [9]. The ACD is claimed to require no phenomenological input, but in one of its versions the locations of the dominant resonances in the spectrum need to be known [10]. Therefore, if the resonances are not known that method depends on two new parameters, although the dependence is not necessarily so strong as to invalidate the entire approach [11].

A new version of of the ACD – called "modified ACD" in this contribution – introduces a weighting function straight into the dispersive integral [12] instead of using it only for the polynomial approximation. Mathematically, this is the most radical departure from the original ACD, which is the main motivation behind this contribution.

2. Original and Modified ACD Method

Finding the matrix element of a process with the vacuum polarization $\Pi(s)$ in the non-perturbative regime may lead to the dispersive integral

$$\int_{s_0}^{\infty} \frac{1}{s}\,\mathrm{Im}F(s)ds,$$

$F(s)$ being usually a simple function of $\Pi(s)$. The lower cut-off s_0 is the threshold of the process. However, $F(s)$ can be related to, e. g., a product of $\Pi(s)$ and another function, as we will see below.

The function $F(s)$ is analytic in the entire complex plane except on the branch cut along the real axis from s_0 so the Cauchy's integral formula gives

$$F(t) = \frac{1}{2\pi i}\oint_C \frac{F(s)}{s-t}ds, \qquad (1)$$

where C is the contour shown in Fig. 1.

The integral consists of two parts [5]:

$$F(t) = \frac{1}{2\pi i}\oint_{|s|=R} \frac{F(s)}{s-t}ds + \frac{1}{\pi}\int_{s_0}^{R} \frac{\mathrm{Im}F(s)}{s-t-i\varepsilon}ds. \qquad (2)$$

In the limit $t \to 0$ Eq. 2 becomes

$$F(0) = \frac{1}{2\pi i}\oint_{|s|=R} \frac{F(s)}{s}ds + \frac{1}{\pi}\int_{s_0}^{R} \frac{\mathrm{Im}F(s)}{s}ds. \qquad (3)$$

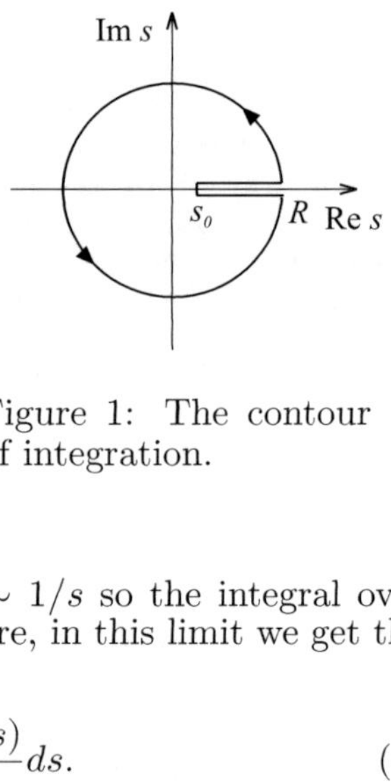

Figure 1: The contour of integration.

It can be shown that asymptotically $F(s) \sim 1/s$ so the integral over the circle $|s| = R$ vanishes as $R \to +\infty$. Therefore, in this limit we get the dispersion relation

$$\overline{F} \equiv F(0) = \frac{1}{\pi}\int_{s_0}^{+\infty} \frac{\mathrm{Im}F(s)}{s}ds. \qquad (4)$$

In the original ACD, the function F is simply related to the vacuum polarization $\Pi(s)$. In the modified ACD we take $F(s) = G(s)\rho(s)$, where $\rho(s)$ is a real continuous function obeying $\rho(0) = 1$ while $G(s)$ is simply related to $\Pi(s)$. Now instead of (3) we find

$$G(0) = \frac{1}{2\pi i} \oint_{|s|=R} \frac{G(s)\rho(s)}{s} ds + \frac{1}{\pi} \int_{s_0}^{R} \frac{\text{Im}G(s)\rho(s)}{s} ds \tag{5}$$

and instead of (4)

$$\overline{G} \equiv G(0) = \frac{1}{\pi} \int_{s_0}^{+\infty} \frac{\text{Im}G(s)\rho(s)}{s} ds. \tag{6}$$

In contrast to the approach that leads to the dispersion relation (4), the ACD method makes the *second* integral in (3), i. e. in (5) for modified ACD, vanish. This is achieved by approximating the kernel $1/s$ by a polynomial. In the original ACD, the kernel is approximated by

$$p_N(s) = \sum_{n=0}^{N} a_n(N) s^n$$

and we find [5]

$$F(0) = \frac{1}{2\pi i} \oint_{|s|=R} \left[\frac{1}{s} - p_N(s)\right] F(s)\, ds + \frac{1}{\pi} \int_{s_0}^{R} \left[\frac{1}{s} - p_N(s)\right] \text{Im}F(s)\, ds. \tag{7}$$

Eq. (7) is exact. We try to make the second integral as small as possible; it is called "fit error". The fit error depends on N, on the interval $[s_0, R]$ and on the fit routine. Clearly, the minimization of the second integral in (5) is more involved as the function $\rho(s)$ is undetermined.

3. Approximation Routine for Modified ACD

The problem of minimization of the fit error is to find the norm that is to be minimized and to find an appropriate function $\rho(s)$. The approach taken in [12] is to divide the weight function into two parts: $\rho(s) = D_1(s)D_2(s)$. The two parts are actually found employing the ideas that have already been used within the original ACD. Since the (usually asymptotic) large-s expansion tends to be better near the upper bound R it makes sense to require that the function $D_2(s)$ and its derivatives vanish at $s = R$:

$$\left.\frac{d^k}{ds^k} D_2(s)\right|_{s=R} = 0, \qquad k = 0, \ldots, N_2. \tag{8}$$

This was the idea of the simplified ACD [6], but in that case there was no (nontrivial) $D_1(s)$ so that version gave only a lower bound of the ACD-estimates. In fact [7],

$$D_2(s) = \left(1 - \frac{s}{R}\right)^{N_2+1}. \tag{9}$$

The function $D_1(s)$ is taken to be the polynomial that minimizes the weighted L_2-norm

$$||d|| = \left[\int_{s_0}^{R} [d(s)]^2 D_2(s) ds \right]^{1/2}, \tag{10}$$

where

$$d(s) = \frac{D_1(s)}{s} = \frac{1}{s} - p_{N_1}(s) = \frac{1}{s} - \sum_{n=0}^{N_1} a_n(N_1, N_2) s^n, \tag{11}$$

With a change of variables $x = -ps + q$, where

$$p = \frac{2}{R - s_0}, \qquad q = \frac{R + s_0}{R - s_0}$$

the approximation problem is reduced to the minimization of the integral

$$||d'||^2 = \int_{-1}^{1} \left[\frac{1}{x - q} - \sum_{n=0}^{N_1} b_n(N_1, N_2) x^n \right]^2 (1 + x)^{N_2+1} dx, \tag{12}$$

where $||d'||$ differs from the norm $||d||$ in (10) only by an overall factor. The change of variables has rescaled and *inverted* the interval: s_0 corresponds to $+1$ and R to -1. There is also an overall factor $-p$:

$$\frac{1}{s} = \frac{-p}{x - q}.$$

It is straightforward to express a_n through b_n [9]:

$$a_n = \sum_{k=n}^{N} \binom{k}{i} (-p)^{n+1} q^{k-n} b_k. \tag{13}$$

Thus the problem is reduced to finding the coefficients b_k.

It is known [13] that the polynomials which minimize the integral (12) are special case of the Jacobi polynomials:

$$y_n(x) = P_n^{(0, N_2+1)}(x), \qquad n = 0, \ldots, N_1$$

and that they are orthogonal with the above weight:

$$\int_{-1}^{1} y_m(x) y_n(x) (1 + x)^{N_2+1} dx = d_n^2 \delta_{mn},$$

where the constant $d_n = d_n(N_2)$ is obtained by specializing the general normalization factor for Jacobi polynomials [14]

$$d_n^2 = \frac{2^{N_2+2}}{2n + N_2 + 2}.$$

Hence we fit the function $1/(x-q)$ to truncated Fourier-Jacobi series:

$$\frac{1}{x-q} \approx \sum_{n=0}^{N_1} c_n y_n(x),$$

where c_n are the Fourier-Jacobi coefficients. We have studied the general case of approximation with Jacobi polynomials in our previous contribution [9] so we need only to specialize the results to the present case. The Fourier-Jacobi coefficients are

$$c_n = \frac{1}{d_n^2} \int_{-1}^{1} \frac{y_n(x)(1+x)^{N_2+1}}{x-q} dx, \qquad q > 1. \tag{14}$$

The integral in (14) for Jacobi polynomials $P_n^{(0,N_2+1)}(x)$ can be expressed in terms of the Jacobi functions of the second kind [13]:

$$Q_n^{(0,N_2+1)}(q) = \frac{1}{2(q+1)^{N_2+1}} \int_{-1}^{1} \frac{P_n^{(0,N_2+1)}(x)(1+x)^{N_2+1}}{q-x} dx. \tag{15}$$

Eventually the coefficients are found to be

$$c_n(q) = -2\frac{Q_n^{(0,N_2+1)}(q)}{d_n^2}(q+1)^{N_2+1}, \qquad n = 0, \ldots, N_1. \tag{16}$$

Jacobi functions of the second kind are easily computable since they are corecursive with the Jacobi polynomials. It is necessary to find Q_0 and Q_1 to initialize the recursive procedure. Actually, the only non-polynomial function that appears in Q_n is the one that is present in Q_0 and it is logarithmic [15]. In our case, the function is

$$\ln\frac{1+q}{1-q} = \ln\frac{R}{s_0}.$$

However, In most realistic applications $N_1 \leq 2$ so that it is not difficult to obtain the fits to $1/s$ directly from (10) and (11).

Finally, we find the function $\rho(s)$:

$$\rho(s) = D_1(s)D_2(s) = \left[1 - \sum_{n=0}^{N_1} a_n(N_1, N_2)s^{n+1}\right]\left(1 - \frac{s}{R}\right)^{N_2+1}. \tag{17}$$

The function $\rho(s)$ is a polynomial of order $N_1 + N_2 + 2$:

$$\rho(s) = \sum_{n=0}^{N_1+N_2+2} A_n(N_1, N_2)s^n, \tag{18}$$

which has been fully determined and enables us to derive the ACD estimate. Of course, not all of the A_n's are independent (e. g. $A_0 = 1$ for all N_1, N_2); in fact, only $N_1 + 1$ A_n's can be considered independent.

4. Derivation of the ACD Estimate

Let us assume that the large-s expansion

$$G(s) = \sum_{m=1}^{M} \frac{h_m(s)}{s^m} + O\left(\frac{1}{s^{M+1}}\right) \tag{19}$$

is valid around the circle $|s| = R$. The expansion (19) may be OPE, which is an asymptotic series, the large-momentum expansion of a perturbative series or an exactly known vacuum polarization, which may be convergent series (the model considered in Sec. 5 belonging to the last case). For the subasymptotic QCD only a few terms of the OPE are known, thus in realistic applications $M \leq 4$.

Returning to the main ACD formula (5) the ACD estimate reeds

$$G(0) = \frac{1}{2\pi i} \oint_{|s|=R} \frac{G(s)\rho(s)}{s} ds + \Delta_{\text{fit}}, \tag{20}$$

where Δ_{fit} is the fit error.

The truncation of the series (19) introduces a further error and the neglect of the momentum dependence of the large-momentum expansion coefficients h_m (which is not always necessary, however) an additional error [5]. We do not consider these two errors here since the fit error is the only error for the model of Sec. 5. The only non-zero contributions to the first integral in (5) come from the simple poles so that

$$G_{\text{ACD}} = \sum_{n=0}^{\min(N_1+N_2+2,M-1)} \hat{h}_{n+1} A_n(N_1, N_2). \tag{21}$$

If the spectral function is known exactly, the dispersion relation gives a reliable estimate of $\overline{G}$ so the comparison to G_{ACD} will determine if the ACD estimate is reliable.

Apparently, the modified ACD introduces two integer parameters N_1 and N_2 as opposed to the original ACD which has only one (the order of the fit polynomial N). However, in most early applications $M \leq 4$ so that looking at (21) we can restrict $N_1 + N_2$ to just 0 or 1. Therefore, there are essentially only three possibilities: 1) $N_1 = N_2 = 0$, 2) $N_1 = 0, N_2 = 1$ and 3) $N_1 = 1, N_2 = 0$. To be fair, one has to note that the actual application in [12] used an expansion with $M = 10$, which allowed to choose among many more N_1, N_2.

5. ACD Estimates for Model Spectra

In order to investigate the reliability of the ACD-derived results we use a simple model function $G(s)$ for which the dispersive integral (6) can be evaluated exactly. Our model contains two infinitely sharp (δ-function) resonances – a single resonance of mass m_V in the vector channel and a single resonance of mass m_A in the axial channel. The parameters of the resonances are constrained by the condition that the spectra saturate the first and second Weinberg sum rules. We find

$$\mathrm{Im}G(s) = -\pi[f_V^2\delta(s-m_V^2) - f_A^2\delta(s-m_A^2)],$$

$$G(s) = \frac{f_V^2}{s-m_V^2+i\epsilon} - \frac{f_A^2}{s-m_A^2+i\epsilon}.$$

Clearly,

$$\overline{G} = G(0) = \frac{f_A^2}{m_A^2} - \frac{f_V^2}{m_V^2} \tag{22}$$

provided $s_0 < m_V, m_A < R$. The Weinberg sum rules give [5]

$$\overline{G} = (1+r)\frac{f^2}{m_V^2}, \tag{23}$$

where $r = m_V^2/m_A^2$ (for QCD $r \approx 0.4$). The large-momentum expansion is

$$G(s) = \sum_{n=0}^{+\infty} \frac{f_V^2 m_V^{2n} - f_A^2 m_A^{2n}}{s^{n+1}} = \sum_{n=0}^{+\infty} \frac{f^2 m_V^{2n}}{s^{n+1}} \frac{1-r^{1-n}}{1-r}.$$

Interestingly, in this model the term with $1/s^2$ is absent.

It is possible to use the more realistic model with finite-width resonances. However, our previous analysis indicated that the model yields quallitatively the same results hence we have not used it in our recent contributions.

In general, only a few terms of the OPE are known [1, 2] so in most applications $M \leq 4$, although there are exceptions. In order to check the sensitivity of the estimates to the location of the resonances we have calculated G_{ACD} not only for its QCD value $r = 0.4$, but also for $r = 0.3$ and $r = 0.5$. This is equivalent to varying the mass m_A of the axial channel resonance with the mass m_V of the vector channel resonance fixed. The results are shown in Table 1.

We have used only the lowest-order approximations with $N_1 + N_2 = 0, 1$. It is clear not only that the results are unsatisfactory, but also that increasing the order of the approximation made the results worse! The computationally much more demanding cases with $N_1+N_2 = 8$ (or so) could perhaps yield better results and will be the subject of further investigation.

Another negative surprise was the strong dependence on the lower cut-off s_0 (not shown on the table), much more severe than in our previous contribution [9]. That dependence is a known problem with the ACD [16, 8] and it is particularly serious for problems with no natural lower cut-off (such as the two-pion threshold) as was the case for the technicolor.

Table 1: $G_{\rm ACD}$ in units of its exact value (23) for 3 values of the dimensionless parameter r. The cutoffs are $s_0 = 0.2m_V^2$ and $R = 5m_V^2$.

r	$N_1 = N_2 = 0$	$N_1 = 0, N_2 = 1$	$N_1 = 1, N_2 = 0$
0.3	0.266	−0.042	−0.489
0.4	0.364	0.077	−0.340
0.5	0.405	0.158	−0.200

6. Conclusions

The ACD is an old method of non-perturbative QCD with several known shortcomings. The modified ACD investigated in this contribution represents a major change from a mathematical point of view. It changes not only the fit routine of the polynomial approximation – which is crucial for the ACD, but also the way the ACD estimates are derived. The modified ACD may also be regarded as a generalization of the ACD method because it introduces two undetermined functions thus opening up new possibilities in the development of the method. In this contribution we have investigated only the concrete application of this concept proposed by the authors of the modified ACD.

The modified ACD weights the polynomial approximation towards the upper cut-off, which is expected to improve the estimates from a physical point of view, namely, the subasymptotic QCD improves with increasing momenta. Since the results are poor, it appears that the weighting is too severe, although we have checked the reliability only for the lowest-order approximations.

The dependence of the ACD-derived estimates on the lower cut-off is also a major problem in the modified ACD, even more severe than in the original ACD.

Hence the modified ACD does not appear to improve the method, but further investigation of this interesting approach is necessary. Firstly, one has to check the reliability and stability of the method for considerably higher orders of the approximation. Secondly, the modification of the method the ACD estimates are derived increases the freedom of choice of the approximation methods and should be further explored.

References

[1] M. A. Shifman, A. I. Vainshtein, and V. I. Zakharov, *Nucl. Phys.* **B147** (1979) 385

[2] N. F. Nasrallah, N. A. Papadopoulos and K. Schilcher, *Phys. Lett.* **B113** (1982) 61; **B126** (1983) 379; *Z. Phys.* **C16** (1983) 323.

[3] I. Caprini and C. Verzegnassi, *Il Nuovo Cim.* **80A** (1984) 187.

[4] Siniša R. Ignjatović, *Analytic continuation by duality method in elementary particle physics*, PhD thesis, University of Cincinnati, 1997, and references therein.

[5] S. R. Ignjatović, in *Proc. IV Summer School in Modern Mathematical Physics*, SFIN (Belgrade) **20A**, 239 (2007).

[6] N. F. Nasrallah, *Phys. Rev.* **D63** (2001) paper 054028.

[7] S. R. Ignjatović, *Fortschr. Phys.* **56** (2008), 452.

[8] S. R. Ignjatović, T. Takeuchi, and L. C. R. Wijewardhana, *Phys. Rev.* **D61** (2000) paper 056006.

[9] S. R. Ignjatović, in *Proc. V Summer School in Modern Mathematical Physics*, SFIN (Belgrade) **22A**, 185 (2009).

[10] C. A. Dominguez, N. F. Nasrallah and K. Schilcher, JHEP (2008) 02 072; C. A. Dominguez, *Mod. Phys. Lett.* **A26**, (2011) 691.

[11] S. R. Ignjatović, *Rom J.. Phys.* **57** (2012), 857.

[12] M. Li and J. Liu, *Int. J. Mod. Phys. E* **21** (2012), paper 1250048.

[13] H. Bateman and A. Erdélyi, *Higher Transcedental Functions* v. 2, Mc-Graw Hill, New York, 1953.

[14] A. F. Nikiforov and V. B. Uvarov, *Special Functions of Mathematical Physics*, Birkhäuser, Boston, 1988.

[15] J. Wimp, P. McCabe and J. N. L. Connor, *J. Comp. Appl. Math.* **82** (1997) 447.

[16] M. Kramer et al., *Phys. Rev.* **D34** (1986) 2127.

Derivation of the trigonometric Gaudin Hamiltonians*

N. Manojlović,†
Departamento de Matemática, F. C. T., Universidade do Algarve
Campus de Gambelas, PT-8005-139 Faro, Portugal

Z. Nagy‡
Grupo de Física Matemática da Universidade de Lisboa,
Av. Prof. Gama Pinto 2, PT-1649-003 Lisboa, Portugal

I. Salom§
Institute of Physics, University of Belgrade
P.O. Box 57, 11080 Belgrade, Serbia

Abstract

Following Sklyanin's proposal in the rational case, we derive the generating function of the Gaudin Hamiltonians in the trigonometric case. Our derivation is based on the quasi-classical expansion of the linear combination of the transfer matrix of the inhomogeneous XXZ Heisenberg spin chain and the central element, the so-called Sklyanin determinant. The corresponding Gaudin Hamiltonians are obtained as the residues of the generating function.

PACS: 02.20.Tw, 03.65.Fd, 05.50+q, 75.10.Jm; MSC 2010: 81R12, 82B23.

1. Introduction

Gaudin models were introduced as interacting spins in a chain [1, 2, 3, 4]. In this approach, these models were obtained as a quasi-classical limit of the integrable quantum chains. Moreover, the Gaudin models were extended to any simple Lie algebra, with arbitrary irreducible representation at each site of the chain [4].

The rational $s\ell(2)$ invariant model was studied in the framework of the quantum inverse scattering method [5]. In his studies, Sklyanin used the $s\ell(2)$ invariant classical r-matrix [5]. A generalization of these results to all cases when skew-symmetric r-matrix satisfies the classical Yang-Baxter equation [6] was relatively straightforward [7, 8]. Therefore, considerable

* Work of I. Salom is supported by the Serbian Ministry of Science and Technological Development under grant number OI 171031.

† e-mail address: nmanoj@ualg.pt

‡ e-mail address: zoltan.nagy@m4x.org

§ e-mail address: isalom@ipb.ac.rs

attention has been devoted to Gaudin models corresponding to the the classical r-matrices of simple Lie algebras [9, 10] and Lie superalgebras[11, 12]. In the case of the $s\ell(2)$ Gaudin system, its relation to Knizhnik-Zamolodchikov equation of conformal field theory [13, 14, 15] or the method of Gauss factorization [16], provided alternative approaches to computation of correlation functions. The non-unitary r-matrices and the corresponding Gaudin models have been studied recently, see [17, 18] and the references therein. In [19] we have derived the generating function of the $s\ell(2)$ Gaudin Hamiltonians with boundary terms. Moreover, we have implemented the algebraic Bethe ansatz, based on the appropriate non-unitary r-matrices and the corresponding linear bracket, obtaining the spectrum of the generating function and the corresponding Bethe equations [19].

Here, following Sklyanin's proposal in the rational case [5, 19], we derive the generating function of the Gaudin Hamiltonians in the trigonometric case. Our derivation is based on the quasi-classical expansion of the linear combination of the transfer matrix of the inhomogeneous XXZ Heisenberg spin chain and the central element, the so-called quantum determinant.

2. Inhomogeneous XXZ Heisenberg spin chain

With the aim deriving the Gaudin Hamiltonians in the trigonometric case, we consider the R-matrix of the XXZ Heisenberg spin chain [20, 21, 22]

$$R(\lambda,\eta)=\begin{pmatrix} \sinh(\lambda+\eta) & 0 & 0 & 0 \\ 0 & \sinh(\lambda) & \sinh(\eta) & 0 \\ 0 & \sinh(\eta) & \sinh(\lambda) & 0 \\ 0 & 0 & 0 & \sinh(\lambda+\eta) \end{pmatrix}. \tag{1}$$

This R-matrix satisfies the Yang-Baxter equation in the space $\mathbb{C}^2\otimes\mathbb{C}^2\otimes\mathbb{C}^2$ [24, 23]

$$R_{12}(\lambda-\mu)R_{13}(\lambda)R_{23}(\mu)=R_{23}(\mu)R_{13}(\lambda)R_{12}(\lambda-\mu). \tag{2}$$

Here we study the inhomogeneous XXZ spin chain with N sites, characterised by the local space $V_m=\mathbb{C}^{2s+1}$ and an inhomogeneous parameter α_m. The Hilbert space of the system is

$$\mathcal{H}=\bigotimes_{m=1}^{N} V_m=(\mathbb{C}^{2s+1})^{\otimes N}. \tag{3}$$

We introduce the Lax operator as the following two-by-two matrix in the auxiliary space $V_0=\mathbb{C}^2$,

$$\mathbb{L}_{0m}(\lambda)=\frac{1}{\sinh(\lambda)}\begin{pmatrix} \sinh\left(\lambda\mathbb{1}_m+\eta S_m^3\right) & \sinh(\eta)S_m^- \\ \sinh(\eta)S_m^+ & \sinh\left(\lambda\mathbb{1}_m-\eta S_m^3\right) \end{pmatrix}. \tag{4}$$

When the quantum space is also a spin $\frac{1}{2}$ representation, the Lax operator becomes the R-matrix,

$$\mathbb{L}_{0m}(\lambda) = \frac{1}{\sinh(\lambda)} R_{0m}(\lambda - \eta/2).$$

Due to the commutation relations (34), it is straightforward to check that the Lax operator satisfies the RLL-relations

$$R_{00'}(\lambda-\mu)\mathbb{L}_{0m}(\lambda-\alpha_m)\mathbb{L}_{0'm}(\mu-\alpha_m) = \mathbb{L}_{0'm}(\mu-\alpha_m)\mathbb{L}_{0m}(\lambda-\alpha_m)R_{00'}(\lambda-\mu). \tag{5}$$

The so-called monodromy matrix

$$T(\lambda) = \mathbb{L}_{0N}(\lambda - \alpha_N) \cdots \mathbb{L}_{01}(\lambda - \alpha_1) \tag{6}$$

is used to describe the system. For simplicity we have omitted the dependence on the quasi-classical parameter η and the inhomogeneous parameters $\{\alpha_j, j = 1, \ldots, N\}$. Notice that $T(\lambda)$ is a two-by-two matrix acting in the auxiliary space $V_0 = \mathbb{C}^2$, whose entries are operators acting in $\mathcal{H}$

$$T(\lambda) = \begin{pmatrix} A(\lambda) & B(\lambda) \\ C(\lambda) & D(\lambda) \end{pmatrix}. \tag{7}$$

From RLL-relations (5) it follows that the monodromy matrix satisfies the RTT-relations

$$R_{00'}(\lambda - \mu)T_0(\lambda)T_{0'}(\mu) = T_{0'}(\mu)T_0(\lambda)R_{00'}(\lambda - \mu). \tag{8}$$

The periodic boundary conditions and the RTT-relations (8) imply that the transfer matrix

$$t(\lambda) = \mathrm{tr}_0 T(\lambda), \tag{9}$$

commute at different values of the spectral parameter,

$$[t(\mu), t(\nu)] = 0. \tag{10}$$

The RTT-relations (8) admit a central element [5]

$$\Delta\left[T(\lambda)\right] = \mathrm{tr}_{00'} P^-_{00'} T_0(\lambda - \eta/2)\, T_{0'}(\lambda + \eta/2), \tag{11}$$

where

$$P^-_{00'} = \frac{-1}{2\sinh(\eta)} R_{00'}(-\eta) = \frac{\mathbb{1} - \mathcal{P}_{00'}}{2}, \tag{12}$$

where $\mathbb{1}$ is the identity and $\mathcal{P}$ is the permutation in $\mathbb{C}^2 \otimes \mathbb{C}^2$. A straightforward calculation shows that $\Delta\left[T(\lambda)\right]$ is a scalar operator

$$\Delta\left[T(\lambda)\right] = \prod_{m=1}^{N} \frac{\sinh\left(\lambda - \alpha_m + \frac{(2s_m+1)\eta}{2}\right)\sinh\left(\lambda - \alpha_m - \frac{(2s_m+1)\eta}{2}\right)}{\sinh\left(\lambda - \alpha_m + \frac{\eta}{2}\right)\sinh\left(\lambda - \alpha_m - \frac{\eta}{2}\right)}, \tag{13}$$

and therefore, it is evidently central,

$$\Big[\Delta\left[T(\lambda)\right]\,,\,T(\nu)\Big] = 0. \tag{14}$$

In the next section we will seek a linear combination of the transfer matrix (9) and the central element (11) whose quasi-classical expansion yields the generating function of the trigonometric Gaudin Hamiltonians in the case when the periodic boundary conditions are imposed.

3. Trigonometric Gaudin model

The Gaudin models were introduce as a quasi-classical limit of the integrable quantum chains [3, 4]. Therefore it is to be expected that the generating function of the trigonometric Gaudin Hamiltonians could be obtained from the quasi-classical expansion of the transfer matrix of the periodic XXZ Heisenberg spin chain. Thus, our first step is to consider the expansion of the monodromy matrix (6) with respect to the quasi-classical parameter η

$$\begin{aligned}
T(\lambda) &= \mathbb{1} + \eta \sum_{m=1}^{N} \frac{\sigma_0^3 \otimes \cosh(\lambda-\alpha_m) S_m^3 + \frac{1}{2}\left(\sigma_0^+ \otimes S_m^- + \sigma_0^- \otimes S_m^+\right)}{\sinh(\lambda-\alpha_m)} + \frac{\eta^2}{2} \mathbb{1}_0 \otimes \sum_{m=1}^{N} \left(S_m^3\right)^2 \\
&+ \frac{\eta^2}{2} \sum_{\substack{n,m=1 \\ n\neq m}}^{N} \frac{\mathbb{1}_0 \otimes \left(\cosh(\lambda-\alpha_m)\cosh(\lambda-\alpha_n)\, S_m^3 S_n^3 + \frac{1}{2}\left(S_m^+ S_n^- + S_m^- S_n^+\right)\right)}{\sinh(\lambda-\alpha_m)\,\sinh(\lambda-\alpha_n)} \\
&+ \frac{\eta^2}{2} \sum_{m=1}^{N} \sum_{n<m}^{N} \frac{\sigma_0^3 \otimes \left(S_m^- S_n^+ - S_m^+ S_n^-\right) + \sigma_0^+ \otimes \left(\cosh(\lambda-\alpha_m) S_m^3 S_n^- - \cosh(\lambda-\alpha_n) S_m^- S_n^3\right)}{2\sinh(\lambda-\alpha_m)\,\sinh(\lambda-\alpha_n)} \\
&+ \frac{\eta^2}{2} \sum_{m=1}^{N} \sum_{n<m}^{N} \frac{\sigma_0^- \otimes \left(\cosh(\lambda-\alpha_n) S_m^+ S_n^3 - \cosh(\lambda-\alpha_m) S_m^3 S_n^-\right)}{2\sinh(\lambda-\alpha_m)\,\sinh(\lambda-\alpha_n)} \\
&+ \frac{\eta^2}{2} \sum_{m=1}^{N} \sum_{n>m}^{N} \frac{\sigma_0^3 \otimes \left(S_n^- S_m^+ - S_n^+ S_m^-\right) + \sigma_0^+ \otimes \left(\cosh(\lambda-\alpha_n) S_n^3 S_m^- - \cosh(\lambda-\alpha_m) S_n^- S_m^3\right)}{2\sinh(\lambda-\alpha_n)\,\sinh(\lambda-\alpha_m)} \\
&+ \frac{\eta^2}{2} \sum_{m=1}^{N} \sum_{n>m}^{N} \frac{\sigma_0^- \otimes \left(\cosh(\lambda-\alpha_m) S_n^+ S_m^3 - \cosh(\lambda-\alpha_n) S_n^3 S_m^-\right)}{2\sinh(\lambda-\alpha_n)\,\sinh(\lambda-\alpha_m)} + \mathcal{O}(\eta^3).
\end{aligned} \tag{15}$$

It is important to notice that the spin operators S_m^α, with $\alpha = +, -, 3$, on the right hand side of (15) satisfy the usual commutation relations

$$[S_m^3, S_n^\pm] = \pm S_m^\pm\, \delta_{mn}, \qquad [S_m^+, S_n^-] = 2S_m^3\, \delta_{mn}. \tag{16}$$

If the Gaudin Lax matrix is defined by

$$L_0(\lambda) = \sum_{m=1}^{N} \frac{\sigma_0^3 \otimes \cosh(\lambda - \alpha_m) S_m^3 + \frac{1}{2}\left(\sigma_0^+ \otimes S_m^- + \sigma_0^- \otimes S_m^+\right)}{\sinh(\lambda - \alpha_m)} \tag{17}$$

and the quasi-classical property of the R-matrix (1) [23]

$$\frac{1}{\sinh(\lambda)} R(\lambda) = \mathbb{1} - \eta r(\lambda) + \mathcal{O}(\eta^2), \tag{18}$$

where

$$r(\lambda) = \frac{-1}{2\sinh(\lambda)} \left(\cosh(\lambda)(\mathbb{1} \otimes \mathbb{1} + \sigma^3 \otimes \sigma^3) + \frac{1}{2} \left(\sigma^+ \otimes \sigma^- + \sigma^- \otimes \sigma^+ \right) \right), \tag{19}$$

is taken into account, then substitution of the expansion (15) into the RTT-relations (8) yields the so-called Sklyanin linear bracket [5]

$$[L_1(\lambda), L_2(\mu)] = [r_{12}(\lambda - \mu), L_1(\lambda) + L_2(\mu)] \,. \tag{20}$$

The classical r-matrix (19) has the unitarity property

$$r_{21}(-\lambda) = -r_{12}(\lambda), \tag{21}$$

and satisfies the classical Yang-Baxter equation [6]

$$[r_{13}(\lambda), r_{23}(\mu)] + [r_{12}(\lambda - \mu), r_{13}(\lambda) + r_{23}(\mu)] = 0. \tag{22}$$

Thus the Sklyanin linear bracket (20) is anti-symmetric and it obeys the Jacobi identity. It follows that the entries of the Lax matrix (17) generate a Lie algebra relevant for the Gaudin model.

Using the expansion (15) it is evident that

$$\begin{aligned} t(\lambda) = 2 + \eta^2 \sum_{m=1}^{N} \Bigg(& \left(S_m^3\right)^2 \\ & + \textstyle\sum_{n \neq m}^{N} \frac{\cosh(\lambda-\alpha_m)\cosh(\lambda-\alpha_n)\, S_m^3 S_n^3 + \frac{1}{2}\left(S_m^+ S_n^- + S_m^- S_n^+\right)}{\sinh(\lambda-\alpha_m)\,\sinh(\lambda-\alpha_n)} \Bigg) + \mathcal{O}(\eta^3). \end{aligned} \tag{23}$$

Analogously, we can expand (11) to obtain

$$
\begin{aligned}
\Delta\left[T(\lambda)\right] &= \mathbb{1} + \eta \operatorname{tr} L(\lambda) + \tfrac{\eta^2}{2}\left(\operatorname{tr}^2 L(\lambda) - \operatorname{tr} L^2(\lambda)\right) \\
&+ \eta^2 \textstyle\sum_{m=1}^{N} \left(\left(S_m^3\right)^2 + \sum_{n\neq m}^{N} \frac{\cosh(\lambda-\alpha_m)\cosh(\lambda-\alpha_n)\, S_m^3 S_n^3 + \frac{1}{2}\left(S_m^+ S_n^- + S_m^- S_n^+\right)}{\sinh(\lambda-\alpha_m)\,\sinh(\lambda-\alpha_n)} \right) \\
&+ \tfrac{\eta^2}{2} \operatorname{tr}_{00'} P_{00'}^- \textstyle\sum_{m=1}^{N} \sum_{m>n}^{N} \left(\frac{\sigma_0^3 \otimes \left(S_m^- S_n^+ - S_m^+ S_n^+\right) + \sigma_0^+ \otimes \left(\cosh(\lambda-\alpha_m) S_m^3 S_n^- - \cosh(\lambda-\alpha_n) S_m^- S_n^3\right)}{2\,\sinh(\lambda-\alpha_m)\,\sinh(\lambda-\alpha_n)} \right. \\
&+ \left. \frac{\sigma_0^- \otimes \left(\cosh(\lambda-\alpha_n) S_m^+ S_n^3 - \cosh(\lambda-\alpha_m) S_m^3 S_n^+\right)}{2\,\sinh(\lambda-\alpha_m)\,\sinh(\lambda-\alpha_n)} \right) \mathbb{1}_{0'} \\
&+ \tfrac{\eta^2}{2} \operatorname{tr}_{00'} P_{00'}^- \textstyle\sum_{m=1}^{N} \sum_{n>m}^{N} \left(\frac{\sigma_0^3 \otimes \left(S_n^- S_m^+ - S_n^+ S_m^+\right) + \sigma_0^+ \otimes \left(\cosh(\lambda-\alpha_n) S_n^3 S_m^- - \cosh(\lambda-\alpha_m) S_n^- S_m^3\right)}{2\,\sinh(\lambda-\alpha_n)\,\sinh(\lambda-\alpha_m)} \right. \\
&+ \left. \frac{\sigma_0^- \otimes \left(\cosh(\lambda-\alpha_m) S_n^+ S_m^3 - \cosh(\lambda-\alpha_n) S_n^3 S_m^+\right)}{2\,\sinh(\lambda-\alpha_n)\,\sinh(\lambda-\alpha_n)} \right) \mathbb{1}_{0'} \\
&+ \tfrac{\eta^2}{2} \operatorname{tr}_{00'} P_{00'}^- \mathbb{1}_0 \textstyle\sum_{m=1}^{N} \sum_{m>n}^{N} \left(\frac{\sigma_{0'}^3 \otimes \left(S_m^- S_n^+ - S_m^+ S_n^+\right) + \sigma_{0'}^+ \otimes \left(\cosh(\lambda-\alpha_m) S_m^3 S_n^- - \cosh(\lambda-\alpha_n) S_m^- S_n^3\right)}{2\,\sinh(\lambda-\alpha_m)\,\sinh(\lambda-\alpha_n)} \right. \\
&+ \left. \frac{\sigma_{0'}^- \otimes \left(\cosh(\lambda-\alpha_n) S_m^+ S_n^3 - \cosh(\lambda-\alpha_m) S_m^3 S_n^+\right)}{2\,\sinh(\lambda-\alpha_m)\,\sinh(\lambda-\alpha_n)} \right) \\
&+ \tfrac{\eta^2}{2} \operatorname{tr}_{00'} P_{00'}^- \mathbb{1}_0 \textstyle\sum_{m=1}^{N} \sum_{n>m}^{N} \left(\frac{\sigma_{0'}^3 \otimes \left(S_n^- S_m^+ - S_n^+ S_m^+\right) + \sigma_{0'}^+ \otimes \left(\cosh(\lambda-\alpha_n) S_n^3 S_m^- - \cosh(\lambda-\alpha_m) S_n^- S_m^3\right)}{2\,\sinh(\lambda-\alpha_n)\,\sinh(\lambda-\alpha_m)} \right. \\
&+ \left. \frac{\sigma_{0'}^- \otimes \left(\cosh(\lambda-\alpha_m) S_n^+ S_m^3 - \cosh(\lambda-\alpha_n) S_n^3 S_m^+\right)}{2\,\sinh(\lambda-\alpha_n)\,\sinh(\lambda-\alpha_n)} \right) + \mathcal{O}(\eta^3),
\end{aligned}
\tag{24}
$$

where $L(\lambda)$ is given in (17). The final expression for the expansion of $\Delta\left[T(\lambda)\right]$ is obtained after taking all the traces

$$
\begin{aligned}
\Delta\left[T(\lambda)\right] &= \mathbb{1} + \eta^2 \\
&\times \sum_{m=1}^{N} \left(\left(S_m^3\right)^2 + \sum_{n\neq m}^{N} \frac{\cosh(\lambda-\alpha_m)\cosh(\lambda-\alpha_n)\, S_m^3 S_n^3 + \frac{1}{2}\left(S_m^+ S_n^- + S_m^- S_n^+\right)}{\sinh(\lambda-\alpha_m)\,\sinh(\lambda-\alpha_n)} \right) \\
&- \frac{\eta^2}{2} \operatorname{tr} L^2(\lambda) + \mathcal{O}(\eta^3).
\end{aligned}
\tag{25}
$$

To obtain the generation function of the Gaudin Hamiltonians notice that (23) and (25) yield

$$
t(\lambda) - \Delta\left[T(\lambda)\right] = \mathbb{1} + \frac{\eta^2}{2} \operatorname{tr} L^2(\lambda) + \mathcal{O}(\eta^3). \tag{26}
$$

Therefore

$$
\tau(\lambda) = \frac{1}{2} \operatorname{tr} L^2(\lambda) \tag{27}
$$

commute for different values of the spectral parameter,

$$
\left[\tau(\lambda), \tau(\mu)\right] = 0. \tag{28}
$$

Moreover, substituting (17) into (27) it is straightforward to obtain the expansion

$$\tau(\lambda) = \sum_{m=1}^{N} \frac{s_m(s_m+1)}{\sinh^2(\lambda - \alpha_m)} + 2\sum_{m=1}^{N} \coth(\lambda - \alpha_m)\, H_m + (S_{gl}^3)^2 , \tag{29}$$

with the Gaudin Hamiltonians

$$H_m = \sum_{n \neq m}^{N} \coth(\alpha_m - \alpha_n)\, S_m^3 S_n^3 + \frac{1}{2\sinh(\alpha_m - \alpha_n)} \left(S_m^+ S_n^- + S_m^- S_n^+\right) \tag{30}$$

and the global generator

$$S_{gl}^3 = \sum_{m=1}^{N} S_m^3. \tag{31}$$

The global generator defined above generates the $U(1)$ symmetry

$$\left[S_{gl}^3, H_m\right] = 0, \quad \text{with} \quad m = 1, 2 \ldots N. \tag{32}$$

Evidently, we have

$$[H_m, H_n] = 0, \quad \text{with} \quad m, n = 1, 2 \ldots N. \tag{33}$$

This shows that $\tau(\lambda)$ is the generating function of Gaudin Hamiltonians (30) when the periodic boundary conditions are imposed [5].

4. Conclusion

Following Sklyanin's proposal [5, 19], we have derive the generating function of the Gaudin Hamiltonians in the trigonometric case by considering the quasi-classical expansion of the linear combination of the transfer matrix of the XXZ Heisenberg spin chain and the corresponding quantum determinant. The Gaudin Hamiltonians are obtained as the residues of the generating function. It would be of considerable interest to generalise these results to the case of non-periodic boundary conditions.

A Basic definitions

We consider the operators S^α with $\alpha = +, -, 3$, acting in some (spin s) representation space $\mathbb{C}^{2s+1}$ with the commutation relations [25]

$$[S^3, S^\pm] = \pm S^\pm, \quad [S^+, S^-] = \frac{\sinh(2\eta S^3)}{\sinh(\eta)} = [2S^3]_q, \tag{34}$$

with $q = e^{\eta}$. In the space $\mathbb{C}^{2s+1}$ these operators admit the following matrix representation [25, 26, 27]

$$S^3 = \sum_{i=1}^{2s+1} a_i e_{i\,i}, \quad S^+ = \sum_{i=1}^{2s+1} b_i e_{i\,i+1}, \quad \text{and} \quad S^- = \sum_{i=1}^{2s+1} b_i e_{i+1\,i} \tag{35}$$

where

$$(e_{ij})_{kl} = \delta_{i\,k}\delta_{j\,l}, \quad a_i = s+1-i, \quad b_i = \sqrt{[i]_q\,[2s+1-i]_q} \quad \text{and} \quad [x]_q = \frac{q^x - q^{-x}}{q - q^{-1}}. \tag{36}$$

In the particular case of spin $\frac{1}{2}$ representation, one recovers the Pauli matrices

$$S^\alpha = \frac{1}{2}\sigma^\alpha = \frac{1}{2}\begin{pmatrix} \delta_{\alpha 3} & 2\delta_{\alpha +} \\ 2\delta_{\alpha -} & -\delta_{\alpha 3} \end{pmatrix}.$$

We consider a spin chain with N sites with spin s representations, i.e. a local $\mathbb{C}^{2s+1}$ space at each site and the operators

$$S_m^\alpha = \mathbb{1} \otimes \cdots \otimes \underbrace{S^\alpha}_{m} \otimes \cdots \otimes \mathbb{1}, \tag{37}$$

with $\alpha = +, -, 3$ and $m = 1, 2, \ldots, N$.

References

[1] M. B. Green, J. H. Schwarz and E. Witten, *Superstring Theory*, Cambridge University Press, 1987.

[2] E. S. Fradkin and A. A. Tseytlin, *Phys.Lett.* **B 158** (1985) 316; *Nucl.Phys.* **B 261** (1985) 1; C. G. Callan, D. Friedan, E. J. Martinec and M. J. Perry, *Nucl.Phys.* **B 262** (1985) 593; T. Banks, D. Nemeschansky and A. Sen, *Nucl.Phys.* **B 277** (1986) 67; A. A. Tseytlin, *Int. J. Mod. Phys.* **A 4** (1989 1257; A. A. Tseytlin, *Int. J. Mod. Phys.* **A 4** (1989 1257.

[3] M. Gaudin, *Diagonalisation d'une classe d'hamiltoniens de spin*, J. Physique **37** (1976) 1087–1098.

[4] M. Gaudin, *La fonction d'onde de Bethe*, chapter 13 Masson, Paris, 1983.

[5] E. K. Sklyanin, *Separation of variables in the Gaudin model*, Zap. Nauchn. Sem. Leningrad. Otdel. Mat. Inst. Steklov. (LOMI) **164** (1987) 151–169; translation in Journal of Soviet Mathematics Volume **47**, Issue 2 (1989) 2473–2488.

[6] A. A. Belavin and V. G. Drinfeld. *Solutions of the classical Yang-Baxter equation for simple Lie algebras* (in Russian), Funktsional. Anal. i Prilozhen. **16** (1982), no. 3, 1–29; translation in Funct. Anal. Appl. **16** (1982) no. 3, 159-180.

[7] E. K. Sklyanin and T. Takebe, *Algebraic Bethe ansatz for the XYZ Gaudin model*, Phys. Lett. **A 219** (1996) 217-225.

[8] M. A. Semenov-Tian-Shansky, *Quantum and classical integrable systems*, in Integrability of Nonlinear Systems, Lecture Notes in Physics **495** (1997) 314-377.

[9] B. Jurčo, *Classical Yang-Baxter equations and quantum integrable systems*, J. Math. Phys. **30** (1989) 1289–1293.

[10] B. Jurčo, *Classical Yang-Baxter equations and quantum integrable systems (Gaudin models)*, in Quantum groups (Clausthal, 1989), Lecture Notes in Phys. **370** (1990) 219–227.

[11] P. P. Kulish and N. Manojlović, *Creation operators and Bethe vectors of the* $osp(1|2)$ *Gaudin model*, J. Math. Phys. **42** no. 10 (2001) 4757–4778.

[12] P. P. Kulish and N. Manojlović, *Trigonometric* $osp(1|2)$ *Gaudin model*, J. Math.Phys. **44** no. 2 (2003) 676–700.

[13] H. M. Babujian and R. Flume, *Off-shell Bethe ansatz equation for Gaudin magnets and solutions of Knizhnik-Zamolodchikov equations*, Mod. Phys. Lett. **A 9** (1994) 2029–2039.

[14] B. Feigin, E. Frenkel, and N. Reshetikhin, *Gaudin model, Bethe ansatz and correlation functions at the critical level*, Commun. Math. Phys. **166** (1994) 27–62.

[15] N. Reshetikhin and A. Varchenko, *Quasiclassical asymptotics of solutions to the KZ equations*, in Geometry, Topology and Physics, Conf. Proc. Lecture Notes Geom., pages 293–273. Internat. Press, Cambridge, MA, 1995.

[16] E. K. Sklyanin, *Generating function of correlators in the* sl_2 *Gaudin model*, Lett. Math. Phys. **47** (1999) 275–292.

[17] T. Skrypnyk, *Non-skew-symmetric classical r-matrix, algebraic Bethe ansatz, and Bardeen-Cooper-Schrieffer-type integrable systems*, J. Math. Phys. **50** (2009) 033540, 28 pages.

[18] T. Skrypnyk, *"Z2-graded" Gaudin models and analytical Bethe ansatz*, Nuclear Physics **B 870** (2013), no. 3, 495–529.

[19] N. Cirilo António, N. Manojlović, E. Ragoucy and I. Salom, *Algebraic Bethe ansatz for the* $s\ell(2)$ *Gaudin model with boundary*, Nuclear Physics **B 893** (2015) 305-331; `arXiv:1412.1396`.

[20] N. Cirilo António, N. Manojlović and Z. Nagy, *Trigonometric* $s\ell(2)$ *Gaudin model with boundary terms*, Reviews in Mathematical Physics **25** No. 10 (2013) 1343004 (14 pages); `arXiv:1303.2481`.

[21] R. J. Baxter, *Partition function of the Eight-Vertex lattice model*, Annals of Physics **70**, Issue 1, March 1972, Pages 193-228.

[22] L. A. Takhtajan and L. D. Faddeev, *The quantum method for the inverse problem and the* XYZ *Heisenberg model*, (in Russian) Uspekhi Mat. Nauk **34** No. 5 (1979) 13–63; translation in Russian Math. Surveys **34** No.5 (1979) 11–68.

[23] P. P. Kulish and E. K. Sklyanin, *Quantum spectral transform method. Recent developments*, Lect. Notes Phys. **151** (1982), 61–119.

[24] R. J. Baxter, *Exactly solved models in statistical mechanics*, Academic Press, London (1982).

[25] P. P. Kulish and N. Yu. Reshetikhin, *Quantum linear problem for the sine-Gordon equation and higher representations*, Zap. Nauchn. Sem. Leningrad. Otdel. Mat. Inst. Steklov. (LOMI) **101** (1981) 101–110, translation in Journal of Soviet Mathematics **23**, Issue 4 (1983) Pages 2435-2441.

[26] L. D. Faddeev, N. Yu. Reshetikhin and L. A. Takhtajan, *Quantum groups*, Braid group, knot theory and statistical mechanics, 97–110, Adv. Ser. Math. Phys., 9, World Sci. Publ., Teaneck, NJ, 1989.

[27] A. Doikou, *A note on the boundary spin s XXZ chain*, Physics Letters **A 366** (2007) 556-562.

Polarization of Photons in Matter-Antimatter Universe*

S.S. Moskaliuk[†]
Bogolyubov Institute for Theoretical Physics,
Metrolohichna Street, 14-b, Kyiv-143, Ukraine, UA-03143

ABSTRACT

In this paper we demonstrate the possibility of generation of linear polarization of the electromagnetic field (EMF) due to the quantum effects of the EMF in matter-antimatter annihilation for anisotropic space of the I type according to Bianchi. The importance of the EMF polarisation concerning the physics of matter-antimatter annihilation in the early Universe relies on the fact that scalar fluctuations and the photon quantum effects can produce only linear polarisation and no circular polarisation. A measurement of the circular polarisation could therefore be interpreted as the detection of gravitational waves caused by matter-antimatter annihilation. Here we consider the generation of the polarization of the EMF in matter-antimatter annihilation process under idealized supposition that the medium is transparent and there is only the CMB radiation before matter-antimatter annihilation. The result of the paper is the assertion that the quantum effects of the EMF in the external gravitational field in the space of the I type according to Bianchi give contribution to the degree of polarization of the EMF in the quadrupole harmonics.

---

* Work was supported in part by the R. M. Santilli Foundation, the Austrian Academy of Sciences and SEENET-MTP project "Cosmology and Strings" PRJ-09 .

† e-mail address: mss@bitp.kiev.ua

1. Polarization of Photons in Matter-Antimatter Annihilation

1.1. Formalism of the EMF polarisation

A useful way to characterise the polarisation properties of the EMF is to use the Stokes parameters formalism [1, 2]. For a nearly monochromatic plane electro-magnetic wave propagating in the z direction,

$$E_x = a_x(t)\cos\left[\omega_0 t - \theta_x(t)\right], E_y = a_y(t)\cos\left[\omega_0 t - \theta_y(t)\right] \tag{1}$$

the Stokes parameters are defined by:

$$\begin{aligned} I \equiv \langle a_x^2\rangle + \langle a_y^2\rangle, \quad Q \equiv \langle a_x^2\rangle - \langle a_y^2\rangle, \\ U \equiv \langle 2a_x a_y \cos(\theta_x - \theta_y)\rangle, \quad V \equiv \langle 2a_x a_y \sin(\theta_x - \theta_y)\rangle, \end{aligned} \tag{2}$$

where the brackets $\langle\rangle$ represent time averages. The parameter I is simply the average intensity of the radiation. The polarisation properties are described by the remaining parameters: Q and U describe linear polarisation, while V describes circular polarisation. Unpolarised radiation (or natural light) is characterised by having $Q = U = V = 0$. The EMF polarisation is produced through the photon quantum effects (see below) which cannot generate circular polarisation. Then, we can write $V = 0$ always for.
The Stokes parameters Q and U are not scalar quantities. If we rotate the reference frame by an angle ϕ around the direction of observation, Q and U transform as:

$$Q' = Q\cos(2\phi) + U\sin(2\phi), U' = U\cos(2\phi) - Q\sin(2\phi). \tag{3}$$

We can define a *polarisation vector* $\mathbf{P}$ having:

$$|\mathbf{P}| = \left(Q^2 + U^2\right)^{1/2}, \qquad \alpha = \frac{1}{2}\tan^{-1}\left(\frac{U}{Q}\right). \tag{4}$$

Although $\mathbf{P}$ is a good way to visualise polarisation, it is not properly a vector, since it remains identical after a rotation by π around z, thus defining an orientation but not a direction. Mathematically, Q and U can be thought as the components of the second-rank symmetric trace-free tensor:

$$\mathbf{P}_{ab} = \frac{1}{2}\begin{pmatrix} Q & -U\sin\theta \\ -U\sin\theta & -Q\sin^2\theta \end{pmatrix}, \tag{5}$$

where the trigonometric functions come from having adopted a spherical coordinate system.

1.2. Formalism of the EMF radiation in anisotropic space of the I type according to Bianchi model

Let us consider Maxwell equations for the free the EMF. In the metrics

$$ds^2 = dt^2 - \sum_{i=1}^{3} A_i^2(t)\left(dx^i\right)^2, \tag{6}$$

they can be written as

$$\nabla_\mu F^{\mu\nu} = 0\,, \qquad \nabla_\mu (*F)^{\mu\nu} = 0\,, \tag{7}$$

$F_{\mu\nu}$ is electro-magnetic-field tensor and $(*F)^{\mu\nu}$ is adjoint magnitude, defined by the relation $(*F)^{\alpha\beta} = \frac{1}{\sqrt{-g}}[\alpha\beta\gamma\eta]F_{\gamma\eta}$, and $[\alpha,\beta,\gamma,\eta]$ is completely antisymmetric tensor [0123]=1.

The solutions of these equations can be represented in the form of electric- and magnetic-field vectors [1]

$$\begin{aligned}
\mathbf{E}(t,\mathbf{x}) &= \int d^3k e^{i\mathbf{kx}} \left[\mathcal{E}^\theta(t,\mathbf{k})\mathbf{e}_\theta(t,\mathbf{k}) + \mathcal{E}^\varphi(t,\mathbf{k})\mathbf{e}_\varphi(t,\mathbf{k})\right], \\
\mathbf{H}(t,\mathbf{x}) &= \int d^3k e^{i\mathbf{kx}} \left[\mathcal{H}^\theta(t,\mathbf{k})\mathbf{e}_\theta(t,\mathbf{k}) + \mathcal{H}^\varphi(t,\mathbf{k})\mathbf{e}_\varphi(t,\mathbf{k})\right],
\end{aligned}$$

where

$$\begin{aligned}
\mathbf{e}_\theta &= \cos\theta_t \cos\varphi_t \frac{\mathbf{e}_1}{A_1} + \cos\theta_t \sin\varphi_t \frac{\mathbf{e}_2}{A_2} - \sin\theta_t \frac{\mathbf{e}_3}{A_3}\,, \\
\mathbf{e}_\varphi &= -\sin\varphi_t \frac{\mathbf{e}_1}{A_1} + \cos\varphi_t \frac{\mathbf{e}_2}{A_2},
\end{aligned}$$

are the orthogonal vectors forming together with

$$\mathbf{e}_k = \sin\theta_t \cos\varphi_t \frac{\mathbf{e}_1}{A_1} + \sin\theta_t \sin\varphi_t \frac{\mathbf{e}_2}{A_2} + \cos\theta_t \frac{\mathbf{e}_3}{A_3}$$

the tetrad unit basis in the momentum space. The angles θ_t and φ_t are related with the spherical coordinates in the momentum space introduced via the relation

$$(k_1, k_2, k_3) = k\,(\sin\theta\,\cos\varphi,\ \sin\theta\,\sin\varphi,\ \cos\theta)\ ,$$

using the formula

$$\begin{aligned}
&(\sin\theta_t\,\cos\varphi_t,\ \sin\theta_t\,\sin\varphi_t,\ \cos\theta_t) = \\
&= \mu^{-1}\left(\frac{\sin\theta\,\cos\varphi}{A_1}, \frac{\sin\theta\,\sin\varphi}{A_2}, \frac{\cos\theta}{A_3}\right),
\end{aligned}$$

where

$$\mu^2 = \frac{\sin^2\theta\,\cos^2\varphi}{A_1^2} + \frac{\sin^2\theta\,\sin^2\varphi}{A_2^2} + \frac{\cos^2\theta}{A_3^2}\,.$$

The components $\mathcal{E}^{\theta}$, $\mathcal{E}^{\varphi}$, $\mathcal{H}^{\vartheta}$, $\mathcal{H}^{\varphi}$ can also be written as follows [2]:

$$\begin{aligned}
\mathcal{E}^{\theta}(t,\mathbf{k}) &= \frac{1}{\sqrt{2}(2\pi)^{3/2}(-g)^{1/4}}\,\frac{\mu}{b^{1/2}}(y^{+}+y^{-})\,,\\
\mathcal{E}^{\varphi}(t,\mathbf{k}) &= \frac{1}{\sqrt{2}(2\pi)^{3/2}(-g)^{1/4}}\,\frac{\mu}{b^{1/2}k}\,\frac{d}{dt}(y^{+}-y^{-})\,,\\
\mathcal{H}^{\theta}(t,\mathbf{k}) &= \frac{1}{\sqrt{2}(2\pi)^{3/2}(-g)^{1/4}}\,\frac{\mu}{b^{1/2}}(y^{+}-y^{-})\,,\\
\mathcal{H}^{\varphi}(t,\mathbf{k}) &= -\frac{1}{\sqrt{2}(2\pi)^{3/2}(-g)^{1/4}}\,\frac{\mu}{b^{1/2}k}\,\frac{d}{dt}(y^{+}+y^{-})\,,
\end{aligned} \tag{8}$$

where

$$b=\frac{1}{\sqrt{-g}}\left(A_2^2\cos^2\varphi+A_1^2\sin^2\varphi\right)\,,$$

and the functions $y^{\pm}=y^{r}$ satisfy the equation

$$\begin{aligned}
\ddot{y}^{r}-\frac{\dot{b}}{b}\dot{y}^{r}+\left[k^2\mu^2+rk\Delta\right]y^{r}=0\,, \qquad (9)\\
\Delta=b\frac{d}{dt}(a/b)\,;\quad a=\frac{\cos\theta\sin 2\varphi}{2\sqrt{-g}}\left(A_2^2-A_1^2\right)\,.
\end{aligned}$$

1.3. Quantum generation of photons in matter-antimatter annihilation

Assuming that at the time moment t_{in} on the background of the initially homogeneous and isotropic gravitational field in the Universe with Friedman metrics there arises the homogeneous anisotropic perturbation in accordance with matter-antimatter annihilation so that as a consequence the metrics can be represented as in Eq. (6). Let us assume also that at $t<t_{\text{in}}$ the state of the EMF can be described with the density matrix with non-zero occupation number of the photons in the mode $n_0(\nu_0)$ corresponding to the black-body radiation. The latter is strictly constant at $t<t_{\text{in}}$ and constant in the zeroth in respect to the anisotropy parameters approximation at $t>t_{\text{in}}$:

$$\frac{\partial}{\partial t}n_0(\nu_0)=0.$$

The frequency ν_0 is considered to be independent of time and equal to the radiation frequency in the current epoch. With the frequency at any time moment t it is related as follows

$$\nu_0 A(t_0)=\nu(t)A(t), \tag{10}$$

where $A(t)$ is the scale factor in the Friedman model at $t < t_{\text{in}}$ and $A^3 = (A_1 A_2 A_3)$ at $t > t_{\text{in}}$.

The external gravitational field of the anisotropic Universe brings about the increase of the photon number [2,4]. The particle number in the mode at the time moment $t > t_{\text{in}}$ satisfies the relation [2]

$$n\,(t,\nu_0,\theta,\varphi) = n_0(\nu_0) + n_1\,(t,\nu_0,\theta,\varphi) + n_q\,(t,\nu_0,\theta,\varphi)\,. \tag{11}$$

Here

$$n_1\,(t,\nu_0,\theta,\varphi) = n_0(\nu_0)\delta\,(t,\nu_0,\theta,\varphi)\,, \tag{12}$$

where $|\delta| \ll 1$ is the correction, describing the anisotropic distribution over momenta which arises due to the anisotropic expansion of the photons already existing to the time moment t_{in}. As to the quantity

$$n_q\,(t,\nu_0,\theta,\varphi) = 2\left(\sum_{r-\pm 1} |\beta^r\,(t,\nu_0,\theta,\varphi)\,|^2\right)(2n_0(\nu_0)+1)\,, \tag{13}$$

it is the additional number of photons which arose due to their generation by matter-antimatter annihilation in the non-stationary gravitational field. β^r is the coefficient of the Bogoliubov transformation at the transition to the time-independent operators of the generation-annihilation of photons bringing to the diagonal form the instantaneous Hamiltonian of the quantised the EMF at the time moment t on the operators of the generation-annihilation, in terms of which the Hamiltonian has the diagonal form at the initial time moment t_{in} ; $|\beta^r(t,\nu_0,\theta,\varphi)|^2$ is the density of the probability of the generation of a photon with a certain frequency ν_0, direction of the wave vector θ,φ, and the spin projection r on the direction of the wave vector.

Let us generalise the relations (11) – (13) to the case when the matter of interest is the particle number in the mode, which polarisation vector is oriented along a certain direction in the coordinate frame connected with the wave vector of the photon. Then by analogy to Eq. (4) we can introduce the symbolic vector [2]

$$\mathbf{n}\,(t,\nu_0,\theta,\varphi) = \frac{1}{2}\frac{c^3}{h\nu^3(t)}\begin{pmatrix} I\,(t,\nu_0,\theta,\varphi) + Q\,(t,\nu_0,\theta,\varphi) \\ I\,(t,\nu_0,\theta,\varphi) - Q\,(t,\nu_0,\theta,\varphi) \\ U\,(t,\nu_0,\theta,\varphi) \end{pmatrix}, \tag{14}$$

where I, Q, U are the Stokes parameters of the EM radiation. By analogy with Eq. (11), $\mathbf{n}$ can be represented as

$$\mathbf{n}\,(t,\nu_0,\theta,\varphi) = \mathbf{n}_0(\nu_0) + \mathbf{n}_1\,(t,\nu_0,\theta,\varphi) + \mathbf{n}_q\,(t,\nu_0,\theta,\varphi)\,, \tag{15}$$

where

$$\mathbf{n}_0(\nu_0) = n_0(\nu_0)\begin{pmatrix} 1 \\ 1 \\ 0 \end{pmatrix},$$

which corresponds to the isotropic nonpolarised radiation. In the case when there is no scattering of the photons on the electrons of the cosmic plasma, $\mathbf{n}_1$ can be represented as [2]

$$\mathbf{n}_1(t,\nu_0,\theta,\varphi) = n_0(\nu_0)\left(\alpha(t,\nu_0)\mathbf{a}(\theta,\varphi) + \bar{\alpha}(t,\nu_0)\bar{\mathbf{a}}(\theta,\varphi)\right),$$

$$\mathbf{a} = \begin{pmatrix} 1 \\ 1 \\ 0 \end{pmatrix}\left(\cos^2\theta - \frac{1}{3}\right), \bar{\mathbf{a}} = \frac{1}{2}\begin{pmatrix} 1 \\ 1 \\ 0 \end{pmatrix}\left(1-\cos^2\theta\right)\cos 2\varphi. \tag{16}$$

This corresponds to the start of the dependence on angles, i.e. anisotropy, in the distribution of the photons over momenta. The coefficients α and $\bar{\alpha}$ characterise the degree of the quadrupole anisotropy of the CMB radiation.

The quantity $\mathbf{n}_q$ describes the contribution of the quantum effects in matter-antimatter annihilation process. In the linear with respect to the anisotropy parameters approximation of the metrics (6) it can be represented in the form:

$$\mathbf{n}_q(t,\nu_0,\theta,\varphi) = \frac{2n\alpha(\nu_0)+1}{2}\frac{c^3}{h\nu^3(t)}\begin{pmatrix} Q^{\text{ann}}(t,\nu_0,\theta,\varphi) \\ -Q^{\text{ann}}(t,\nu_0,\theta,\varphi) \\ 2U^{\text{ann}}(t,\nu_0,\theta,\varphi) \end{pmatrix}. \tag{17}$$

The quantities I^{ann}, Q^{ann}, U^{ann} are the annihilation Stokes parameters (ASP), evaluated for the case when the initial state of the EMF at the time of the matter-antimatter annihilation t_{in} was only the CMB radiation.

2. Polarization of Photons due to the quantum effects in matter-antimatter Universe

In general (i.e. not in the linear approximation) the Stokes parameters are related with the polarisation density matrix of the quantum effects of the EMF in matter-antimatter Universe, introduced in [1] as a new characteristics of the latter, as usually (see [2]):

$$J^{ab} = \frac{1}{2}\begin{pmatrix} I^{\text{ann}} + Q^{\text{ann}} & U^{\text{ann}} - iV^{\text{ann}} \\ U^{\text{ann}} + iV^{\text{ann}} & I^{\text{ann}} - Q^{\text{ann}} \end{pmatrix}, \tag{18}$$

where

$$J^{ab\,ann} = J_{+}^{ab\,ann} + J_{-}^{ab\,ann}, \quad J_{\pm}^{ab\,ann}(t,\mathbf{k}) = \frac{1}{2}\frac{hk}{\mu(t,\theta,\varphi)} \times$$
$$\times \langle 0_{t_{\text{in}}}|N_t\left[\hat{\mathcal{E}}^a(t,\mathbf{x},\mathbf{k}),\hat{\mathcal{E}}^b(t,\mathbf{x},\mathbf{k})\right]|0_{t_{\text{in}}}\rangle.$$

Here $\hat{\mathcal{E}}^a(t,\mathbf{x},\mathbf{k})$ are the components of the spectral component of the vector of electric field (8), multiplied by $\exp(i\mathbf{kx})$.

The calculations, carried out in [2], have shown that $J^{ab\,ann} = 0$, i.e. according to the common interpretation [3], the generating photons do not have an admixture of the circular polarisation.

The symmetric part of the polarisation tensor could be expressed via spectral components of the averages of the operator of energy-momentum tensor

$$T_{\mu\nu}^{\text{ann}}(t) = \int d\varphi d\theta \sin\theta \int dK_0\left(t,k,\theta,\varphi\right) \widetilde{T}_{\mu\nu}^{\text{ann}}\left(t,k,\theta,\varphi\right),$$
$$K_0\left(t,k,\theta,\varphi\right) = ck\mu\left(t,\theta,\varphi\right),$$

as follows

$$J_+^{ab} = G_{\mu\nu}^{ab}\widetilde{T}_{\mu\nu}^{\text{ann}}.$$

The specific form of the components $G_{\mu\nu}^{ab}$ has been evaluated in [5]. Also in [2, 5] is given the explicit form of the spectral components $\tilde{T}_{\mu\nu}^{\text{ann}}$. So the ASP can be presented as follows [2] :

$$\left(I^{\text{ann}}, Q^{\text{ann}}, U^{\text{ann}}, V^{\text{ann}}\right) = \frac{hk^3}{V}\sum_{r=\pm 1}\left(2s^r, u^r, r\tau^r, 0\right), \tag{19}$$

where the functions s^r, u^r, τ^r satisfy the set of the equations [2]

$$\begin{cases} \dot{s}^r = \frac{W}{2}u^r + r\frac{\overline{W}}{2}\tau^r, \\ \dot{u}^r = W\left(2s^r+1\right) - \left(r\overline{W} + 2cK_0\right)\tau^r, \\ \dot{\tau}^r = r\overline{W}\left(2s^r+1\right) + \left(r\overline{W} + 2cK_0\right)u^r, \end{cases} \tag{20}$$
$$\left(u^r\right)^2 + \left(\tau^r\right)^2 = 4s^r\left(s^r+1\right)$$

with the initial values $s^r(t_{\text{in}}) = u^r(t_{\text{in}}) = \tau^r(t_{\text{in}}) = 0$. The quantities W and $\overline{W}$ in the metrics, linear in respect to the anisotropy parameters of metrics (6) are as follows:

$$\begin{aligned} W &= \left(1-\cos^2\theta\right)\Delta H + \frac{\overline{\Delta H}}{2}\left(1+\cos^2\theta\right)\cos 2\varphi, \\ \overline{W} &= -\cos\theta\,\sin 2\varphi\overline{\Delta H}, \end{aligned} \tag{21}$$

where

$$\begin{aligned} \Delta H &= = H - \frac{1}{2}\left(H_1 + H_2\right), \quad \overline{\Delta H} = H_1 - H_2, \\ H^3 &= \left(H_1 H_2 H_3\right), \quad H_i = \dot{A}_i/A_i \end{aligned}$$

(the parameters ΔA, $\overline{\Delta A}$, A could be introduced similarly).

The set of the equations (20) plays the part of the transfer equations for ASP. Let us analyse the expressions (19), solving (20) via expansion of the functions in a power series in respect to small parameter $\tilde{h}$ which is introduced as

$$\Delta H \to \tilde{h}\Delta H, \quad \overline{\Delta H} \to \tilde{h}\overline{\Delta H}, \Delta A \to \tilde{h}\Delta A, \quad \overline{\Delta A} \to \tilde{h}\overline{\Delta A}.$$

In doing so,

$$s^r = \sum_{n=0} \tilde{h}^n s_n^r, \quad u^r = \sum_{n=0} \tilde{h}^n u_n^r, \quad \tau^r = \sum_{n=0} \tilde{h}^n \tau_n^r. \tag{22}$$

The expansions of W and $\overline{W}$ are given with (21). Further we shall keep in the expansion (22) only linear terms in respect to $\tilde{h}$. In the zeroth approximation in respect to $\tilde{h}$ it follows from (20), (21), taking into the account the initial values, that

$$s_0^r = u_0^r = \tau_0^r = 0.$$

This means that there are no photon quantum effects in isotropic case. In the linear approximation the set of the equations for s_1^r, u_1^r, τ_1^r is

$$\dot{s}_1^r = 0, \quad \dot{u}_1^r = W - 2\nu\tau_1^r, \quad \dot{\tau}_1^r = r\overline{W} + 2\nu u_1^r$$

(in the zeroth approximation in respect to $\tilde{h}$ $K_0(t, k, \theta, \varphi) = k_0/A(t) \equiv \nu(t)$). Solving this set, one can obtain the expressions for ASP in the linear approximation:

$$(I^{\text{ann}}, Q^{\text{ann}}, U^{\text{ann}}) = \tag{23}$$

$$= \frac{2h\nu^3}{c^3} \int_{t_{\text{in}}}^{t} \left(0, W(t'), \overline{W}(t')\right) \cos\left(2(\Omega(t) - \Omega(t'))\right) dt',$$

$$\Omega(t) = \int dt \nu(t).$$

Such a distribution of the ASP quantum effects in the anisotropic gravitational field is unusual from the viewpoint of the classical electrodynamics [6]. The reason is the structure of the vacuum energy-momentum tensor of the EMF in the external gravitational field. *Zeldovich* and *Starobinsky* [7] have remarked that quantum effects of the material field in the external anisotropic gravitational field bring about the breaking of the condition of the energy dominance of the energy-momentum tensor (EMT) of these fields.) This fact shows itself in different dependence of the EMT components on the anisotropy parameters of the metrics (6), namely:

$$T_0^0 \sim \tilde{h}^2, \qquad T^{ik} \sim \tilde{h}, \qquad i, k = 1, 2, 3.$$

The fact that EMT does not satisfy the condition of the energy dominance means that it contains both the contribution from the really generated particles and the contribution of the EMF in matter-antimatter annihilation process. It is impossible to divide the energy-momentum tensor into the aforementioned contributions as it has been indicated in [2].

Let us bring $\mathbf{n}_q$ to the form analogous that of $\mathbf{n}_1$, singling out explicitly the dependence on the angles θ and φ. To this end we use (21) and (23), then

$$\begin{aligned} \mathbf{n}_q(t, \nu_0, \theta, \varphi) &= \frac{2n_0(\nu_0)+1}{2} \times \\ &\times \left(\beta_q(t,\nu_0)\mathbf{b}(\theta,\varphi) + \bar{\beta}_q(t,\nu_0)\bar{\mathbf{b}}(\theta,\varphi)\right), \end{aligned} \tag{24}$$

here

$$\beta_q(t) = \int_{t_{\text{in}}}^{t} \Delta H(t') \cos\left(2(\Omega(t) - \Omega(t'))\right) dt'. \tag{25}$$

The expression for $\bar{\beta}_q$ can be obtained from (25), substituting $\overline{\Delta H}$ for ΔH. Vectors $\mathbf{b}$ and $\bar{\mathbf{b}}$ are defined via the relations (analogously as was defined in [8] by Basco and Polnarev):

$$\mathbf{b} = \begin{pmatrix} 1 \\ -1 \\ 0 \end{pmatrix} (1-\cos^2\theta), \quad \bar{\mathbf{b}} = \frac{1}{2} \begin{pmatrix} (1+\cos^2\theta)\cos 2\varphi \\ -(1+\cos^2\theta)\cos 2\varphi \\ 4\cos\theta\,\sin 2\varphi \end{pmatrix}.$$

Assuming that $n_0 \gg 1$ and substituting (16), (24) into (15), we obtain

$$\mathbf{n} = \mathbf{n}_0 + n_0\left[\alpha\mathbf{a} + \beta_q\mathbf{b} + \bar{\alpha}\bar{\mathbf{a}} + \bar{\beta}_q\bar{\mathbf{b}}\right]. \tag{26}$$

The quantities β_q and $\bar{\beta}_q$ are related with the degree of the linear polarisation of the EMF in matter-antimatter annihilation. The relation (26) is similar to that derived in [8] under the solution the radiation transfer equation taking into the account Thomson scattering of the photons on the electrons of the cosmic plasma.

The quantities which undergo the measurement in the course of experiment are the Stokes parameters I, Q and U. Let us evaluate, for example, the Stokes parameter Q in the Heckman-Schüking model [2]. According to Eqs. (14) and (26) we have

$$Q(t,\nu_0,\theta,\varphi) = \frac{2h\nu_0^3}{c^3} n_0(\nu_0)\left(1-\cos^2\theta\right)\beta_q(t,\nu_0).$$

The dependence of Q on the time t is determined with the quantity β_q (25). Let us transform it to more clear form. To this end let us temporarily change in the integral (25) the integration variable X as follows:

$$\begin{aligned} X &= (1+\chi)^{-1/2}, \quad \Delta H = \Delta H_0 (1+\chi)^3 = \Delta H_0 / X^6, \\ dt &= -\frac{d\chi}{H_0(1+\chi)^{5/2}} = \frac{2}{H_0} X^2 dX, \nu(t) = \nu_0(1+\chi) = \nu_0 X^{-2} \end{aligned}$$

where ΔH_0, H_0, ν_0 are the current values of the corresponding parameters. By this change of variable the integral is brought to the form

$$\beta_q = \frac{\Delta H_0}{H_0} \int\limits_{X_{\mathrm{in}}}^{X} \frac{\cos \lambda (X - X')}{X'^4} dX', \tag{27}$$

where $\lambda = 4\nu_0 / H_0 \approx 10^{30}$ is a large parameter. Estimating (27) asymptotically in λ, we obtain

$$\beta_q = \frac{\Delta H_0}{H_0} \frac{1}{\lambda X_{\mathrm{in}}^4} \sin\left(\lambda (X - X_{\mathrm{in}})\right). \tag{28}$$

Let us change in the last formula from the current variable X to the synchronous time t according to the relation $X = (3H_0 t/2)^{1/3}$. The time t is synchronous cosmological time, counted from the singularity. Let us divide it into two summands as follows:

$$t = t_0 + t', \qquad t' \ll t_0,$$

where $t_0 = \frac{2}{3H_0}$ is the time counted from the beginning of the expansion, corresponding to the current epoch, and t' is the current time, for example, the period, during which the observations take place. Then

$$\begin{aligned} Q(t, \nu_0, \theta, \varphi) &= \frac{2h\nu_0^3}{c^3} n_0(\nu_0)(1 - \cos^2\theta) \times \\ &\times \frac{\Delta H_0}{H_0} \frac{\chi_{\mathrm{in}}}{\lambda} \sin\left(2\nu_0 t' + \lambda(1 - X_{\mathrm{in}})\right). \end{aligned}$$

Let us discuss the possibility of experimental measuring of Q. The power of the polarised component of interest for us of the EMF radiation, hitting the aerial of the radio-telescope, with the directivity diagram $P_n(\theta, \varphi)$ and the effective area of the surface A_{eff} per unit frequency band is defined as follows [9]:

$$W(t) = \frac{1}{2} A_{\mathrm{eff}} \left| \iint\limits_{\Omega} Q(t, \nu_0, \theta, \varphi) P_n(\theta, \varphi) d\Omega \right|. \tag{29}$$

Separating the dependence on angle, we bring (29) to the form

$$W(t) = \frac{2h\nu_0^3}{c^3} n_0(\nu_0) \Omega_{\mathrm{eff}}^{Q} \left| \beta_q(t, \nu_0) \right|.$$

The quantity $W(t)$ is the instantaneous power of the signal in the aerial. On the load of the aerial the voltage is induced, the square of which is proportional to the instantaneous power, so that the current at the output of the detector is as follows [9]:

$$i^Q_{\text{det}}(t) = k'\hat{U}^2(t) = k'W(t) = k'\Omega^Q_{\text{eff}}\,\frac{2h\nu_0^3}{c^3}n_0(\nu_0)\left|\beta_q(t,\nu_0)\right| .$$

After averaging over time, we obtain

$$\overline{i^Q_{\text{det}}} = k'\left(\Omega^Q_{\text{eff}}\right)\left(\frac{2h\nu_0^3}{c^3}n_0(\nu_0)\right)\frac{\Delta H_0}{H_0}\frac{\chi^2_{\text{in}}}{\lambda}\frac{2}{\pi} .$$

Similarly, measuring the Stokes parameter I in the zeroth approximation with respect to the anisotropy parameter ΔH, we arrive at

$$\overline{i^I_{\text{det}}} = k'\left(\Omega^T_{\text{eff}}\right)\left(\frac{2h\nu_0^3}{c^3}n_0(\nu_0)\right) .$$

The observable degree of polarisation turns out to be

$$P = \frac{\overline{i^Q_{\text{det}}}}{\overline{i^I_{\text{det}}}} = \frac{2}{\pi}\frac{\Omega^Q_{\text{eff}}}{\Omega^I_{\text{eff}}}\frac{\Delta H_0}{H_0}\frac{\chi^2_{\text{in}}}{\lambda} . \qquad (30)$$

Thus, the degree of polarisation of the EMF due to the quantum effects in matter-antimatter process turns out to be of essence only in the case, when the anisotropy of the metrics has manifested itself sufficiently early.

The background result of this paper constitutes the idea, which attempts to explain some linear polarization of the EMF in the Universe as produced by quantum effects in matter-antimatter annihilation process. Thus, it turns out that in the case of the transparent medium when there is no scattering of the photons, the EMF radiation in the homogeneous anisotropic and non-stationary Universe becomes linearly polarised due to the quantum effects of the photon generation by matter-antimatter annihilation. It is remarkable that the angular dependence of the photon number is quadrupole and completely coincides with that arising under the scattering of the CMB photons on electrons in the epoch of the recombination or the secondary ionization [1–4]. On the other hand, the generation of the photons in matter-antimatter annihilation process and the polarization of the EMF take place in the external anisotropic gravitational field in the space of the I type according to Bianchi [1–4].

References

[1] S. S. Moskaliuk, and et al., *Phys. At. Nucl.* **58**, (1995) 1744–1747.

[2] S. Moskaliuk, *Phys. Rev. D* **68**, (2003) 084023.

[3] S. S. Moskaliuk, "Propagation of Polarized Cosmic Microwave Background Radiation in an Anisotropic Magnetized Plasma", in *Proceedings of the 4-th Gamow International Conference on Astrophysics and Cosmology after Gamow and The 9th Gamow Summer School (17-23 August, 2009, Odessa, Ukraine)*, edited by S. K. C. et al., AIP Conference Proceedings 1206, American Institute of Physics, New York, 2009, pp. 64–78.

[4] S. Moskaliuk, *Ukrainian Journ. of Phys.* **5**, (2010) 636–643.

[5] A. Sagnotti, and B. Zwiebach, *Phys. Rev. D* **24**, (1981) 305.

[6] V. N. Lukash, and I. D. Novikov, *The very Early Universe*, Cambridge Univ. Press, Cambridge, 1983.

[7] Y. B. Zeldovich, and A. Starobinsky, *Sov. Phys. JETP* **34**, No.6, (1972) 1159–1166.

[8] M. M. Basco, and A. Polnarev, *Astron. Journ.* **57**, No.3, (1980) 465.

[9] J. D. Kraus, *Radio astronomy*, McGraw-Hill, New York, 1966.

[10] S. Ahmad, and et al., *Phys. Lett.* **152**, (1985) 135.

[11] L. Adiels, and et al., *Phys. Lett.* **182**, B (1986) 405.

[12] D. Fixsen, and et al., *Astrophys.J.* **473**, (1996) 576.

[13] Y. Zeldovich, and R. Sunyaev, *Ap. Space Sci.* **4**, (1966) 301.

Creation operators of the non-periodic $s\ell(2)$ Gaudin model*

I. Salom†
Institute of Physics, University of Belgrade
P.O. Box 57, 11080 Belgrade, Serbia
N. Manojlović,‡
Departamento de Matemática, F. C. T., Universidade do Algarve
Campus de Gambelas, PT-8005-139 Faro, Portugal

Abstract

We define new creation operators relevant for implementation of the algebraic Bethe ansatz for the $s\ell(2)$ Gaudin model with the general reflection matrix. This approach is based on the linear bracket corresponding to the relevant non-unitary classical r-matrix.
PACS: 02.20.Tw, 03.65.Fd, 05.50+q, 75.10.Jm; MSC 2010: 81R12, 82B23.

1. Introduction

In [1] we have derived the generating function of the $s\ell(2)$ Gaudin Hamiltonians with boundary terms. We have shown that the implementation of the algebraic Bethe ansatz requires an appropriate non-unitary r-matrices and the corresponding linear bracket [1]. The non-unitary r-matrices and the corresponding Gaudin models have been studied recently, see [2, 3] and the references therein. In [1] we have obtained the spectrum of the generating function and the corresponding Bethe equations. However, explicit and compact form of the Bethe vector $\varphi_M(\mu_1, \mu_2, \ldots, \mu_M)$, for an arbitrary positive integer M, remained open. Our aim here is to propose creation operators which should solve this problem.

2. $s\ell(2)$ Gaudin model with boundary terms

The classical r-matrix relevant for the $s\ell(2)$ Gaudin model is given by [4]

$$r(\lambda) = -\frac{\mathcal{P}}{\lambda}, \tag{1}$$

* Work of I. Salom is supported by the Serbian Ministry of Science and Technological Development under grant number OI 171031.
† e-mail address: isalom@ipb.ac.rs
‡ e-mail address: nmanoj@ualg.pt

where $\mathcal{P}$ is the permutation matrix in $\mathbb{C}^2 \otimes \mathbb{C}^2$. This classical r-matrix satisfies the classical Yang-Baxter equation

$$[r_{13}(\lambda), r_{23}(\mu)] + [r_{12}(\lambda - \mu), r_{13}(\lambda) + r_{23}(\mu)] = 0, \tag{2}$$

and it has the unitarity property

$$r_{21}(-\lambda) = -r_{12}(\lambda). \tag{3}$$

The general solution of the corresponding classical reflection equation [5, 6, 7]:

$$\begin{aligned} & r_{12}(\lambda - \mu) K_1(\lambda) K_2(\mu) + K_1(\lambda) r_{21}(\lambda + \mu) K_2(\mu) = \\ & = K_2(\mu) r_{12}(\lambda + \mu) K_1(\lambda) + K_2(\mu) K_1(\lambda) r_{21}(\lambda - \mu), \end{aligned} \tag{4}$$

is give by [1]

$$\widetilde{K}(\lambda) = \begin{pmatrix} \xi - \lambda & \widetilde{\psi}\lambda \\ \widetilde{\phi}\lambda & \xi + \lambda \end{pmatrix}. \tag{5}$$

An important preliminary step in the implementation of the algebraic Bethe ansatz for the open Gaudin model is to bring the K-matrix (5) to the upper, or lower, triangular form [1]

$$K(\lambda) = U^{-1} \widetilde{K}(\lambda) U = \begin{pmatrix} \xi - \lambda\nu & \lambda\psi \\ 0 & \xi + \lambda\nu \end{pmatrix}, \tag{6}$$

where $\psi = \widetilde{\phi} + \widetilde{\psi}$ and

$$U = \begin{pmatrix} -1 - \nu & \widetilde{\phi} \\ \widetilde{\phi} & -1 - \nu \end{pmatrix}, \tag{7}$$

with $\nu = \sqrt{1 + \widetilde{\phi}\widetilde{\psi}}$.

Here we study the $s\ell(2)$ Gaudin model with N sites, characterised by the local space $V_m = \mathbb{C}^{2s+1}$ and inhomogeneous parameter α_m. The Hilbert space of the system is

$$\mathcal{H} = \underset{m=1}{\overset{N}{\otimes}} V_m = (\mathbb{C}^{2s+1})^{\otimes N}. \tag{8}$$

Following [1] we introduce the Lax operator

$$\mathcal{L}_0(\lambda) = \begin{pmatrix} H(\lambda) & F(\lambda) \\ E(\lambda) & -H(\lambda) \end{pmatrix} = \sum_{m=1}^{N} \left(\frac{\vec{\sigma}_0 \cdot \vec{S}_m}{\lambda - \alpha_m} + \frac{K_0(\lambda) \vec{\sigma}_0 K_0^{-1}(\lambda) \cdot \vec{S}_m}{\lambda + \alpha_m} \right), \tag{9}$$

with the following local realisation for the entries of the Lax matrix

$$E(\lambda) = \sum_{m=1}^{N} \left(\frac{S_m^+}{\lambda - \alpha_m} + \frac{(\xi + \lambda\nu) S_m^+}{(\xi - \lambda\nu)(\lambda + \alpha_m)} \right), \tag{10}$$

$$F(\lambda) = \sum_{m=1}^{N} \left(\frac{S_m^-}{\lambda - \alpha_m} + \frac{(\xi - \lambda\nu)^2 S_m^- - \lambda^2\psi^2 S_m^+ - 2\lambda\psi(\xi - \lambda\nu) S_m^3}{(\xi + \lambda\nu)(\xi - \lambda\nu)(\lambda + \alpha_m)} \right), \tag{11}$$

$$H(\lambda) = \sum_{m=1}^{N} \left(\frac{S_m^3}{\lambda - \alpha_m} + \frac{\lambda\psi\, S_m^+ + (\xi - \lambda\nu) S_m^3}{(\xi - \lambda\nu)(\lambda + \alpha_m)} \right). \tag{12}$$

Due to the commutation relations (36), it is straightforward to check that the Lax operator (9) satisfies the following linear bracket relations

$$[\mathcal{L}_0(\lambda), \mathcal{L}_{0'}(\mu)] = \left[r_{00'}^K(\lambda,\mu), \mathcal{L}_0(\lambda)\right] - \left[r_{0'0}^K(\mu,\lambda), \mathcal{L}_{0'}(\mu)\right], \tag{13}$$

where the non-unitary r-matrix is give by

$$r_{00'}^K(\lambda,\mu) = r_{00'}(\lambda - \mu) - K_{0'}(\mu) r_{00'}(\lambda + \mu) K_{0'}^{-1}(\mu). \tag{14}$$

The commutator (13) is obviously anti-symmetric. It obeys the Jacobi identity because the r-matrix (14) satisfies the classical Yang-Baxter equation

$$[r_{32}^K(\lambda_3,\lambda_2), r_{13}^K(\lambda_1,\lambda_3)] + [r_{12}^K(\lambda_1,\lambda_2), r_{13}^K(\lambda_1,\lambda_3) + r_{23}^K(\lambda_2,\lambda_3)] = 0. \tag{15}$$

The linear bracket (13) based on the r-matrix $r_{00'}^K(\lambda,\mu)$ (14), corresponding to (6) and the classical r-matrix (1), defines the Lie algebra relevant for the open $s\ell(2)$ Gaudin model.

As it was shown in [1], it is instructive to introduce the new generators $e(\lambda), h(\lambda)$ and $f(\lambda)$ as the following linear combinations of the original ones

$$e(\lambda) = \tfrac{-\xi+\lambda\nu}{\lambda} E(\lambda), \quad h(\lambda) = \tfrac{1}{\lambda}\left(H(\lambda) - \tfrac{\psi\lambda}{2\xi} E(\lambda)\right),$$
$$f(\lambda) = \tfrac{1}{\lambda}\left((\xi + \lambda\nu)F(\lambda) + \psi\lambda H(\lambda)\right). \tag{16}$$

The key observation is that in the new basis we have

$$[e(\lambda), e(\mu)] = [h(\lambda), h(\mu)] = [f(\lambda), f(\mu)] = 0. \tag{17}$$

Therefore there are only three nontrivial relations

$$[h(\lambda), e(\mu)] = \frac{2}{\lambda^2 - \mu^2} \left(e(\mu) - e(\lambda) \right), \tag{18}$$

$$\begin{aligned}[h(\lambda), f(\mu)] &= \frac{-2}{\lambda^2 - \mu^2} \left(f(\mu) - f(\lambda) \right) - \frac{2\psi\nu}{(\lambda^2 - \mu^2)\xi} \left(\mu^2 h(\mu) - \lambda^2 h(\lambda) \right) \\ &- \frac{\psi^2}{(\lambda^2 - \mu^2)\xi^2} \left(\mu^2 e(\mu) - \lambda^2 e(\lambda) \right), \end{aligned} \tag{19}$$

$$\begin{aligned}[e(\lambda), f(\mu)] &= \frac{2\psi\nu}{(\lambda^2 - \mu^2)\xi} \left(\mu^2 e(\mu) - \lambda^2 e(\lambda) \right) \\ &- \frac{4}{\lambda^2 - \mu^2} \left((\xi^2 - \mu^2\nu^2) h(\mu) - (\xi^2 - \lambda^2\nu^2) h(\lambda) \right). \end{aligned} \tag{20}$$

In the Hilbert space $\mathcal{H}$ (8), in every $V_m = \mathbb{C}^{2s+1}$ there exists a vector $\omega_m \in V_m$ such that

$$S_m^3 \omega_m = s_m \omega_m \quad \text{and} \quad S_m^+ \omega_m = 0. \tag{21}$$

We define a vector Ω_+ to be

$$\Omega_+ = \omega_1 \otimes \cdots \otimes \omega_N \in \mathcal{H}. \tag{22}$$

From the definitions above, the formulas (10) - (12) and (16) it is straightforward to obtain the action of the generators $e(\lambda)$ and $h(\lambda)$ on the vector Ω_+

$$e(\lambda)\Omega_+ = 0, \quad \text{and} \quad h(\lambda)\Omega_+ = \rho(\lambda)\Omega_+, \tag{23}$$

with

$$\rho(\lambda) = \frac{1}{\lambda} \sum_{m=1}^{N} \left(\frac{s_m}{\lambda - \alpha_m} + \frac{s_m}{\lambda + \alpha_m} \right) = \sum_{m=1}^{N} \frac{2s_m}{\lambda^2 - \alpha_m^2}. \tag{24}$$

The generating function of the Gaudin Hamiltonians with boundary terms is given by [1]:

$$\begin{aligned}\tau(\lambda) = \operatorname{tr}_0 \mathcal{L}_0^2(\lambda) &= 2\lambda^2 \left(h^2(\lambda) + \frac{2\nu^2}{\xi^2 - \lambda^2\nu^2} h(\lambda) - \frac{h'(\lambda)}{\lambda} \right) \\ &- \frac{2\lambda^2}{\xi^2 - \lambda^2\nu^2} \left(f(\lambda) + \frac{\psi\lambda^2\nu}{\xi} h(\lambda) + \frac{\psi^2\lambda^2}{4\xi^2} e(\lambda) - \frac{\psi\nu}{\xi} \right) e(\lambda). \end{aligned} \tag{25}$$

An important initial observation in the implementation of the algebraic Bethe ansatz is that the vector Ω_+ (22) is an eigenvector of the generating function $\tau(\lambda)$. To show this we use the expression (23) and (24):

$$\tau(\lambda)\Omega_+ = \chi_0(\lambda)\Omega_+ = 2\lambda^2 \left(\rho^2(\lambda) + \frac{2\nu^2 \rho(\lambda)}{\xi^2 - \lambda^2\nu^2} - \frac{\rho'(\lambda)}{\lambda} \right) \Omega_+. \tag{26}$$

With the aim of obtaining the explicit and compact form of the Bethe vectors we define the following creation operaters

$$c(\lambda) = f(\lambda) + \frac{\psi\xi}{\nu}\, h(\lambda) + \frac{\psi^2}{4\nu^2}\, e(\lambda). \tag{27}$$

Using the relations (17) - (20) it is straightforward to check that

$$[c(\lambda), c(\mu)] = 0. \tag{28}$$

Consequently, the Bethe vectors generated by the action of the operators (27) on the vector Ω_+ (22) will be symmetric functions of theirs arguments.

Our main aim is to show that the Bethe vector $\varphi_1(\mu)$ has the form

$$\varphi_1(\mu) = c(\mu)\Omega_+ = \left(f(\lambda) + \frac{\psi\xi}{\nu}\, \rho(\lambda) \right) \Omega_+, \tag{29}$$

where $c(\mu)$ is given by (27). The action of the generating function of the Gaudin Hamiltonians reads

$$\tau(\lambda)\varphi_1(\mu) = [\tau(\lambda), c(\mu)]\, \Omega_+ + \chi_0(\lambda)\varphi_1(\mu). \tag{30}$$

Using (25) and the commutation relations (17) - (20) it is evident that

$$[\tau(\lambda), c(\mu)]\, \Omega_+ = [\tau(\lambda), f(\mu)]\, \Omega_+. \tag{31}$$

Then, a straightforward calculation show that

$$\begin{aligned} [\tau(\lambda), f(\mu)]\, \Omega_+ &= -\frac{8\lambda^2}{\lambda^2 - \mu^2} \left(\rho(\lambda) + \frac{\nu^2}{\xi^2 - \lambda^2\nu^2} \right) \varphi_1(\mu) \\ &+ \frac{8\lambda^2(\xi^2 - \mu^2\nu^2)}{(\lambda^2 - \mu^2)(\xi^2 - \lambda^2\nu^2)} \left(\rho(\mu) + \frac{\nu^2}{\xi^2 - \mu^2\nu^2} \right) \varphi_1(\lambda). \end{aligned} \tag{32}$$

Therefore the action of the generating function $\tau(\lambda)$ on $\varphi_1(\mu)$ is given by

$$\tau(\lambda)\varphi_1(\mu) = \chi_1(\lambda, \mu)\varphi_1(\mu) + \frac{8\lambda^2(\xi^2 - \mu^2\nu^2)}{(\lambda^2 - \mu^2)(\xi^2 - \lambda^2\nu^2)} \left(\rho(\mu) + \frac{\nu^2}{\xi^2 - \mu^2\nu^2} \right) \varphi_1(\lambda), \tag{33}$$

with

$$\chi_1(\lambda, \mu) = \chi_0(\lambda) - \frac{8\lambda^2}{\lambda^2 - \mu^2} \left(\rho(\lambda) + \frac{\nu^2}{\xi^2 - \lambda^2\nu^2} \right). \tag{34}$$

The unwanted term in (33) vanishes when the following Bethe equation is imposed on the parameter μ,

$$\rho(\mu) + \frac{\nu^2}{\xi^2 - \mu^2\nu^2} = 0. \tag{35}$$

Thus we have shown that $\varphi_1(\mu)$ (29) is the desired Bethe vector of the generating function $\tau(\lambda)$ corresponding to the eigenvalue $\chi_1(\lambda, \mu)$.

3. Conclusion

We have proposed a new creation operators relevant for implementation of the algebraic Bethe ansatz for the $s\ell(2)$ Gaudin model with the general reflection matrix. However, explicit and compact form of the relevant Bethe vector $\varphi_M(\mu_1, \mu_2, \ldots, \mu_M)$, for an arbitrary positive integer M, requires further studies and will be reported elsewhere. Such a formula would be crucial for the off shell scalar product of the Bethe vectors and these results could lead to the correlations functions of Gaudin model with boundary. Moreover, it would be of considerable interest to establish a relation between Bethe vectors and solutions of the corresponding Knizhnik-Zamolodchikov equations.

A Basic definitions

We consider the spin operators S^α with $\alpha = +, -, 3$, acting in some (spin s) representation space $\mathbb{C}^{2s+1}$ with the commutation relations

$$[S^3, S^\pm] = \pm S^\pm, \quad [S^+, S^-] = 2S^3, \tag{36}$$

and Casimir operator

$$c_2 = (S^3)^2 + \frac{1}{2}(S^+S^- + S^-S^+) = (S^3)^2 + S^3 + S^-S^+ = \vec{S} \cdot \vec{S}.$$

In the particular case of spin $\frac{1}{2}$ representation, one recovers the Pauli matrices

$$S^\alpha = \frac{1}{2}\sigma^\alpha = \frac{1}{2}\begin{pmatrix} \delta_{\alpha 3} & 2\delta_{\alpha +} \\ 2\delta_{\alpha -} & -\delta_{\alpha 3} \end{pmatrix}.$$

We consider a spin chain with N sites with spin s representations, i.e. a local $\mathbb{C}^{2s+1}$ space at each site and the operators

$$S_m^\alpha = \mathbb{1} \otimes \cdots \otimes \underbrace{S^\alpha}_{m} \otimes \cdots \otimes \mathbb{1}, \tag{37}$$

with $\alpha = +, -, 3$ and $m = 1, 2, \ldots, N$.

References

[1] N. Cirilo António, N. Manojlović, E. Ragoucy and I. Salom, *Algebraic Bethe ansatz for the $s\ell(2)$ Gaudin model with boundary*, Nuclear Physics **B 893** (2015) 305-331; `arXiv:1412.1396`.

[2] T. Skrypnyk, *Non-skew-symmetric classical r-matrix, algebraic Bethe ansatz, and Bardeen-Cooper-Schrieffer-type integrable systems*, J. Math. Phys. **50** (2009) 033540, 28 pages.

[3] T. Skrypnyk, *"Z2-graded" Gaudin models and analytical Bethe ansatz*, Nuclear Physics **B 870** (2013), no. 3, 495–529.

[4] E. K. Sklyanin, *Separation of variables in the Gaudin model*, Zap. Nauchn. Sem. Leningrad. Otdel. Mat. Inst. Steklov. (LOMI) **164** (1987) 151–169; translation in Journal of Soviet Mathematics **47**, Issue 2 (1989) 2473–2488.

[5] E. K. Sklyanin, *Boundary conditions for integrable equations*, (Russian) Funktsional. Anal. i Prilozhen. **21** (1987) 86–87; translation in Functional Analysis and Its Applications **21**, Issue 2 (1987) 164–166.

[6] E. K. Sklyanin, *Boundary conditions for integrable systems*, in the Proceedings of the VIIIth international congress on mathematical physics (Marseille, 1986), World Sci. Publishing, Singapore, (1987) 402–408.

[7] E. K. Sklyanin, *Boundary conditions for integrable quantum systems*, J. Phys. A: Math. Gen. **21** (1988) 2375–2389.

Conformal invariance for scalar and Dirac particles in Riemannian spacetimes: New results

Alexander J. Silenko*
Research Institute for Nuclear Problems, Belarusian State University, Minsk 220030, Belarus
BLTP, Joint Institute for Nuclear Research, Dubna 141980, Russia

Abstract

Conformal symmetry properties of pointlike scalar and Dirac particles (Higgs boson and all leptons) in Riemannian spacetimes in the presence of electromagnetic interactions are considered. A Hermitian form of the Klein-Gordon equation for a pointlike scalar particle in an arbitrary n-dimensional Riemannian spacetime is obtained. New conformal symmetries of initial and Hermitian forms of this equation are ascertained. In the above spacetime, general Hamiltonians in the generalized Feshbach-Villars and Foldy-Wouthuysen representations are derived. The conformal-like transformation conserving these Hamiltonians is found. Corresponding conformal symmetries of a Dirac particle are determined. It is proven that all conformal symmetries remain unchanged by an inclusion of electromagnetic interactions.

1. Introduction

A determination of symmetry properties of elementary particles is one of the most important problems of contemporary particle physics. Symmetries of basic relativistic wave equations describing pointlike particles with spin 0 (Higgs boson) and 1/2 (all leptons) retain an important place among these properties. Intensive studies of such symmetries have been started fifty years ago from the seminal work by Penrose [1]. He has discovered the conformal invariance of the covariant Klein-Gordon (KG) equation [2] for a massless scalar particle in a Riemannian spacetime and has supplemented this equation with a term describing a nonminimal coupling to the scalar curvature. Chernikov and Tagirov [3] have involved the case of a nonzero mass and n-dimensional Riemannian spacetime. The inclusion of the additional Penrose-Chernikov-Tagirov (PCT) term has been argued for both massive and massless particles [3]. Accioly and Blas [4] have performed the exact Foldy-Wouthuysen (FW) transformation for a massive spin-0 particle in static spacetimes and have found new telling arguments in favor of the predicted coupling to the scalar curvature. A derivation of the relativistic FW Hamiltonian is very important for a comparison of gravitational (and

* e-mail address: alsilenko@mail.ru

inertial) effects in classical and quantum gravity because the FW representation restores Schrödinger-like forms of Hamiltonians and equations of motion. These forms are convenient for finding a semiclassical approximation and a classical limit of relativistic quantum mechanics (see Refs. [5, 6, 7] and references therein).

However, the transformation method used in Ref. [4] is applicable neither to massless particles nor to nonstatic spacetimes. To find a specific manifestation of the conformal invariance in the FW representation which takes place just for massless particles, the generalized Feshbach-Villars (GFV) transformation [8] applicable for such particles has been performed [9]. The subsequent relativistic FW transformations has made it possible to derive the FW Hamiltonians for the both massive and massless scalar particles in general noninertial frames and stationary gravitational fields. The new manifestation of the conformal invariance for massless particles consisting in the conservation of the FW Hamiltonian and the FW wave function has been discovered. New exact FW Hamiltonians have been obtained for both massive and massless scalar particles in general static spacetimes and in frames rotating in the Kerr field approximated by a spatially isotropic metric. The high precision expression for the FW Hamiltonian has been derived in the general case. It has been also shown that conformal transformations change only such terms in the FW Hamiltonians which are proportional to the particle mass m.

In the present work, we consider the much more general problem of scalar and Dirac particles in arbitrary gravitational (noninertial) and electromagnetic fields and find (on a quantum-mechanical level) new symmetry properties relative to conformal transformations not only in the FW representation but also in initial representations. These properties are attributed to all known pointlike scalar and Dirac particles (Higgs boson and leptons) and also to the hypothetic pseudoscalar axion.

We denote world and spatial indices by Greek and Latin letters, respectively. The signature is $(+---)$, the Ricci scalar curvature is defined by $R = g^{\mu\nu}R_{\mu\nu} = g^{\mu\nu}R^{\alpha}{}_{\mu\alpha\nu}$, where $R^{\alpha}{}_{\mu\beta\nu} = \partial_\beta\Gamma^{\alpha}{}_{\mu\nu} - \ldots$ is the Riemann curvature tensor. We use the system of units $\hbar = 1,\ c = 1$.

2. Hermitian form of the Klein-Gordon equation and conformal symmetry for a pointlike scalar particle

The covariant KG equation with the additional PCT term [1, 3] describing a scalar particle in a n-dimensional Riemannian spacetime is given by

$$(\Box + m^2 - \lambda R)\psi = 0, \quad \Box \equiv \frac{1}{\sqrt{-g}}\partial_\mu\sqrt{-g}g^{\mu\nu}\partial_\nu. \tag{1}$$

Minimal (zero) coupling corresponds to $\lambda = 0$, while the PCT coupling is defined by $\lambda = (n-2)/[4(n-1)]$ [3]. The sign of the Penrose-Chernikov-Tagirov term depends on the definition of R. For noninertial (accelerated and rotating) frames, the spacetime is flat and $R = 0$.

For a *massless* particle, the conformal transformation

$$\widetilde{g}_{\mu\nu} = O^{-2}g_{\mu\nu} \tag{2}$$

conserves the form of Eq. (1) but changes the wave function and the operators acting on it [1, 3]:

$$\Box - \lambda R = O^{-\frac{n+2}{2}}(\widetilde{\Box} - \lambda\widetilde{R})O^{\frac{n-2}{2}}, \quad \widetilde{\psi} = O^{\frac{n-2}{2}}\psi. \tag{3}$$

To specify symmetry properties of the initial KG equation (1), it is instructive to present it in the Hermitian form. Amazingly, this can be achieved with the simple nonunitary transformation

$$\psi = f^{-1}\Phi, \quad f = \sqrt{g^{00}\sqrt{-g}}, \quad g = \det g_{\mu\nu}. \tag{4}$$

Since $\widetilde{g} = O^{-2n}g$, Φ is invariant relative to the conformal transformation (2). After the transformation (4), we multiply the obtained equation by the factor f/g^{00} and come to the Hermitian form of the KG equation [10]:

$$\left(\frac{1}{f}\partial_\mu\sqrt{-g}g^{\mu\nu}\partial_\nu\frac{1}{f} + \frac{m^2}{g^{00}} - \frac{\lambda R}{g^{00}}\right)\Phi = 0. \tag{5}$$

The use of Eqs. (2)–(4) shows that Eq. (5) is conformally invariant for a massless particle. However, it is not conformally invariant for a massive one. To determine its conformal symmetry in the latter case, it is sufficient to find a physical quantity which substitution for m restores the conformal invariance of Eq. (5). For this purpose, we can use the quantity m' which is equal to m in the initial spacetime and takes the form

$$\widetilde{m'} = Om' \tag{6}$$

after the conformal transformation (2). The equation obtained from Eq. (5) with the substitution of m' for m,

$$\left(\frac{1}{f}\partial_\mu\sqrt{-g}g^{\mu\nu}\partial_\nu\frac{1}{f} + \frac{m'^2}{g^{00}} - \frac{\lambda R}{g^{00}}\right)\Phi = 0, \tag{7}$$

is conformally invariant. While this equation does not describe a real particle and is not equivalent to Eq. (5), finding the appropriate substitution (6) determines the conformal symmetry of the suitable equation (5). The determination of new symmetry property for massive particles is rather important because the only discovered pointlike scalar particle, Higgs boson, is massive.

Thus, we can conclude that Eq. (5) is not changed by the conformal-like transformation

$$\widetilde{g}_{\mu\nu} = O^{-2}g_{\mu\nu}, \quad m \to m', \quad \widetilde{m'} = Om'. \tag{8}$$

In the general case, we can substitute any quantity satisfying Eq. (6) for m into Eq. (5).

We can now state the conformal symmetry of the initial KG equation (1). The substitution of m' for m makes the *changed* equation to be conformally invariant with the following properties:

$$\Box + m'^2 - \lambda R = O^{-\frac{n+2}{2}}(\widetilde{\Box} + \widetilde{m'}^2 - \lambda \widetilde{R})O^{\frac{n-2}{2}}, \quad \widetilde{\psi} = O^{\frac{n-2}{2}}\psi. \tag{9}$$

These properties establish the conformal symmetry of the covariant KG equation (1) and the specific form of its invariance relative to the conformal-like transformation (8).

The method of the FW transformation used in Ref. [9] is applicable to nonstationary spacetimes. However, only the stationary case has been considered in this work. To make a more general investigation of symmetry properties in the FW representation, we need to present Eq. (5) in another (equivalent) form.

Let us introduce the following denotations:

$$\Gamma^i = \sqrt{-g}g^{0i}, \quad G^{ij} = g^{ij} - \frac{g^{0i}g^{0j}}{g^{00}}. \tag{10}$$

Lengthy but straightforward calculations bring Eq. (5) to the form [10]

$$\left[(\partial_0 + \Upsilon)^2 + \partial_i \frac{G^{ij}}{g^{00}}\partial_j + \frac{m^2}{g^{00}} + \Lambda\right]\Phi = 0, \tag{11}$$

where

$$\begin{array}{c}
\Upsilon = \frac{1}{2f}\left\{\partial_i, \Gamma^i\right\}\frac{1}{f} = \frac{1}{2}\left\{\partial_i, \frac{g^{0i}}{g^{00}}\right\}, \\
\Lambda = -\frac{f_{,0,0}}{f} - \left(\frac{g^{0i}}{g^{00}}\right)_{,i}\frac{f_{,0}}{f} - 2\frac{g^{0i}}{g^{00}}\frac{f_{,0,i}}{f} - \left(\frac{g^{0i}}{g^{00}}\right)_{,0}\frac{f_{,i}}{f} \\
-\frac{1}{2}\left(\frac{g^{0i}}{g^{00}}\right)_{,0,i} - \frac{1}{2f^2}\left(\frac{g^{0i}}{g^{00}}\right)_{,i}\Gamma^j_{,j} - \frac{g^{0i}}{2f^2g^{00}}\Gamma^j_{,j,i} \\
+\frac{1}{4f^2}\left(\Gamma^i_{,i}\right)^2 - \left(\frac{G^{ij}}{g^{00}}\right)_{,i}\frac{f_{,j}}{f} - \frac{G^{ij}}{g^{00}}\frac{f_{,i,j}}{f} - \frac{\lambda R}{g^{00}}.
\end{array} \tag{12}$$

This form of the KG equation is also Hermitian and the wave function is not changed as compared with Eq. (7). The replacement of m with m' makes Eq. (11) to be conformally invariant. Therefore, Eq. (11) is invariant relative to the conformal-like transformation (8).

3. Conformal symmetries of Hamiltonians

To fulfil the successive GFV and Foldy-Wouthyusen transformations, we use the method developed in Ref. [8] and applied to the covariant KG equation in Ref. [9]. The original Feshbach-Villars method does not work for massless particles while its generalization [8] makes it possible to extend the method on such particles.

We introduce two new functions, ϕ and χ, as follows [8, 9]:

$$\Phi = \phi + \chi, \quad i(\partial_0 + \Upsilon)\Phi = N(\phi - \chi), \tag{13}$$

where N is an arbitrary nonzero real parameter. For the Feshbach-Villars transformation, it is definite and equal to the particle mass m. These functions form the two-component wave function in the GFV representation, $\Psi = \begin{pmatrix} \phi \\ \chi \end{pmatrix}$. Equations (11) and (13) result in (cf. Ref. [9])

$$\begin{array}{c} i\frac{\partial\Psi}{\partial t} = \mathcal{H}\Psi, \quad \mathcal{H} = \rho_3\frac{N^2+T}{2N} + i\rho_2\frac{-N^2+T}{2N} - i\Upsilon, \\ T = \partial_i\frac{G^{ij}}{g^{00}}\partial_j + \frac{m^2}{g^{00}} + \Lambda, \end{array} \tag{14}$$

where $\mathcal{H}$ is the GFV Hamiltonian and ρ_i $(i = 1, 2, 3)$ are the Pauli matrices. Equation (14) is exact.

For a massless particle, this Hamiltonian is not changed by the conformal transformation (2). In the general case, it is invariant relative to the conformal-like transformation (8).

The general methods developed in Refs. [7, 8, 6] allow us to perform the FW transformation of the Hamiltonian (14) for a relativistic particle in external fields. These methods are iterative. The initial Hamiltonian can be presented in the general form

$$\mathcal{H} = \rho_3\mathcal{M} + \mathcal{E} + \mathcal{O}, \quad \rho_3\mathcal{M} = \mathcal{M}\rho_3, \quad \rho_3\mathcal{E} = \mathcal{E}\rho_3, \quad \rho_3\mathcal{O} = -\mathcal{O}\rho_3, \tag{15}$$

where $\mathcal{E}$ and $\mathcal{O}$ denote the sums of even (diagonal) and odd (off-diagonal) operators, respectively. In the considered case, $[\mathcal{M}, \mathcal{O}] = 0$,

$$\mathcal{M} = \frac{N^2+T}{2N}, \quad \mathcal{E} = -i\Upsilon, \quad \mathcal{O} = i\rho_2\frac{-N^2+T}{2N}, \tag{16}$$

and the transformation operator found in Ref. [7] reduces to the form [8, 9]

$$U = \frac{\epsilon + N + \rho_1(\epsilon - N)}{2\sqrt{\epsilon N}}, \quad \epsilon = \sqrt{\mathcal{M}^2 + \mathcal{O}^2} = \sqrt{T}. \tag{17}$$

This operator is ρ_3-pseudounitary ($U^\dagger = \rho_3 U^{-1}\rho_3$).

It is important that the Hamiltonian obtained as a result of this transformation does not depend on N [8]:

$$\begin{array}{c} \mathcal{H}' = \rho_3\epsilon + \mathcal{E}' + \mathcal{O}', \quad \rho_3\mathcal{E}' = \mathcal{E}'\rho_3, \quad \rho_3\mathcal{O}' = -\mathcal{O}'\rho_3, \\ \mathcal{E}' = -i\Upsilon + \frac{1}{2\sqrt{\epsilon}}\left[\sqrt{\epsilon}, \left[\sqrt{\epsilon}, \mathcal{F}\right]\right]\frac{1}{\sqrt{\epsilon}}, \\ \mathcal{O}' = \rho_1\frac{1}{2\sqrt{\epsilon}}[\epsilon, \mathcal{F}]\frac{1}{\sqrt{\epsilon}}, \quad \mathcal{F} = -i\partial_0 - i\Upsilon. \end{array} \tag{18}$$

This shows a self-consistency of the used transformation method. The *exact* intermediate Hamiltonian (18) describes massive and massless particles and is not changed by the conformal-like transformation (8).

Next transformation [8] eliminates residual odd terms and leads to the final form of the *approximate* relativistic FW Hamiltonian:

$$\mathcal{H}_{FW} = \rho_3\epsilon - i\Upsilon - \frac{1}{2\sqrt{\epsilon}}\left[\sqrt{\epsilon}, \left[\sqrt{\epsilon}, (i\partial_0 + i\Upsilon)\right]\right]\frac{1}{\sqrt{\epsilon}}. \tag{19}$$

This final Hamiltonian is also invariant relative to the conformal-like transformation (8). As a rule, the relativistic FW Hamiltonian is expanded in powers of the Planck constant and it is useful when the de Broglie wavelength is much smaller than the characteristic distance [7]. In such a Hamiltonian, terms proportional to the zero and first powers of the Planck constant are determined exactly while less order terms are not specified (see Ref. [11]). As a result, the last term in Eq. (19) can be omitted if it is proportional to the second or higher orders of $\hbar$.

4. Inclusion of electromagnetic interactions

Fortunately, an inclusion of electromagnetic interactions does not leads to any significant complication of the above derivations. The initial covariant KG equation takes the form

$$\left[g^{\mu\nu}(\nabla_\mu + ieA_\mu)(\nabla_\nu + ieA_\nu) + m^2 - \lambda R\right]\psi = 0, \tag{20}$$

where ∇_μ is the covariant derivative and A_μ is the electromagnetic field potential. This equation is equivalent to the following one:

$$\left(\frac{1}{\sqrt{-g}}D_\mu\sqrt{-g}g^{\mu\nu}D_\nu + m^2 - \lambda R\right)\psi = 0, \tag{21}$$

where $D_\mu = \partial_\mu + ieA_\mu$. The nonunitary transformation (4) brings it to the Hermitian form corresponding to Eq. (5) [10]:

$$\left(\frac{1}{f}D_\mu\sqrt{-g}g^{\mu\nu}D_\nu\frac{1}{f} + \frac{m^2}{g^{00}} - \frac{\lambda R}{g^{00}}\right)\Phi = 0. \tag{22}$$

It is convenient to present this equation in the equivalent form (cf. Eq. (11)) [10]

$$\begin{gathered}\left[(D_0 + \Upsilon')^2 + D_i\tfrac{G^{ij}}{g^{00}}D_j + \tfrac{m^2}{g^{00}} + \Lambda\right]\Phi = 0,\\ \Upsilon' = \tfrac{1}{2f}\left\{\partial_i, \Gamma^i\right\}\tfrac{1}{f} = \tfrac{1}{2}\left\{D_i, \tfrac{g^{0i}}{g^{00}}\right\},\\ T' = D_i\tfrac{G^{ij}}{g^{00}}D_j + \tfrac{m^2}{g^{00}} + \Lambda,\end{gathered} \tag{23}$$

where G^{ij} and Λ are defined by Eqs. (10) and (12), respectively.

A repeat of the transformation given above allows us to derive the Hamiltonian in the GFV representation:

$$\mathcal{H} = \rho_3\frac{N^2+T'}{2N} + \rho_2\frac{-N^2+T'}{2N} - i\Upsilon' + eA_0. \tag{24}$$

The FW transformation can be fulfilled with the operator (17) where $\epsilon = \sqrt{T'}$. The transformed operator $\mathcal{H}'$ is independent of N. The final approximate FW Hamiltonian is given by ($\epsilon = \sqrt{T'}$)

$$\mathcal{H}_{FW} = \rho_3\epsilon - i\Upsilon' + eA_0 - \tfrac{1}{2\sqrt{\epsilon}}\left[\sqrt{\epsilon}, \left[\sqrt{\epsilon}, (i\partial_0 + i\Upsilon' - eA_0)\right]\right]\tfrac{1}{\sqrt{\epsilon}}. \tag{25}$$

The last term in Eq. (25) can be omitted if it is proportional to the second or higher orders of $\hbar$ (see previous section).

All Hamiltonians obtained with the inclusion of electromagnetic interactions ($\mathcal{H}$, $\mathcal{H}'$, and $\mathcal{H}_{FW}$) are invariant relative to the conformal-like transformation (8). The Hamiltonians are conformally invariant for a massless particle. Thus, this inclusion does not change the conformal symmetries of the Hamiltonians.

5. Conformal symmetry properties of Dirac particles

It is easier to determine the conformal symmetry properties of a pointlike Dirac particle. It has been established in Ref. [9] that the Dirac and FW Hamiltonians for a *massless* particle and the corresponding wave functions are invariant relative to the conformal transformation (2). The initial covariant Dirac equation is also conformally invariant for such a particle. The wave function of the massless particle in the conformally transformed metric (2) acquires the additional factor $O^{3/2}$ [9].

These results can be extended on massive particles. An analysis of the general relativistic equation for the Dirac Hamiltonian in arbitrary Riemannian spacetimes in the presence of an electromagnetic field (Eq. (2.21) in Ref. [12]) shows that this Hamiltonian is invariant relative to the conformal-like transformation (8). The FW transformation operator possesses the same property for both massive and massless particles. As a result, the FW Hamiltonian is also invariant relative to the conformal-like transformation in the general case. Finally, we find that the conformal-like transformation (8) of the covariant Dirac equation results in the following property of the wave function:

$$\widetilde{\Psi} = O^{3/2}\Psi. \qquad (26)$$

Contrary to the conventional conformal invariance, this property is valid for any particle.

All properties stated in this section take place in the presence of electromagnetic interactions.

We can conclude that the previously ascertained similarity between massless scalar and Dirac particles in Riemannian spacetimes [9] exists for any pointlike particles and is not violated by electromagnetic interactions.

6. Summary

In the present work, new symmetry properties have been found for fundamental pointlike scalar and Dirac particles (Higgs boson and all leptons) in Riemannian spacetimes. All general results have been obtained in the presence of electromagnetic interactions. The KG equation for a pointlike scalar particle in an arbitrary n-dimensional Riemannian spacetime has been brought to the Hermitian form (5). This form is useful to derive the general Hamiltonians in the GFV and FW representations. The GFV Hamiltonians (14) and (24) are exact. The corresponding FW Hamiltonians (19) and (25) are approximate. They are expanded in powers of the

Planck constant and are useful when the de Broglie wavelength is much smaller than the characteristic distance. Nevertheless, these Hamiltonians are rather general. They cover the nonstationary case and can be applied for a relativistic particle in arbitrarily strong gravitational and inertial fields. In these Hamiltonians, terms proportional to the zero and first powers of the Planck constant are determined exactly while less order terms are not specified.

New conformal symmetries of the initial and Hermitian forms of the KG equation are ascertained. When the mass is replaced with any quantity m' satisfying the conformal transformation (6), the *changed* equations become conformally invariant. This defines the conformal symmetries of the conventional and Hermitian KG equations. The latter equation as well as the obtained Hamiltonians in the GFV and FW representations is invariant relative to the conformal-like transformation (8).

Corresponding conformal symmetries are also determined for both massive and massless Dirac particles. The Dirac and FW Hamiltonians are invariant relative to the conformal-like transformation (8). This transformation also defines the conformal symmetry of the initial Dirac equation for a massive particle. When m' defined by Eq. (6) is substituted for m, the Dirac wave function has the property (26).

It is proven that all conformal symmetries remain unchanged by an inclusion of electromagnetic interactions. Thus, the results obtained in the present study have allowed us to state the new general properties of conformal symmetry for pointlike scalar and Dirac particles (Higgs boson and all leptons) in Riemannian spacetimes in the presence of electromagnetic interactions.

Acknowledgements

The author thanks the Organizing Committee of the 8th Mathematical Physics Meeting for hospitality and support. The work was also supported in part by the Belarusian Republican Foundation for Fundamental Research (Grant No. Φ14D-007) and by the Heisenberg-Landau program of the German Ministry for Science and Technology (Bundesministerium für Bildung und Forschung).

References

[1] R. Penrose, in *Relativity, Groups and Topology*, edited by C. DeWitt and B. DeWitt, Gordon and Breach, 1964, p. 565.

[2] O. Klein, Z. Phys. **37** (1926) 895; W. Gordon, Z. Phys. **40** (1926) 117. The equation has been first obtained by E. Schroedinger (unpublished) and also by V. Fock, Z. Phys. **38** (1926) 242.

[3] N. Chernikov and E. Tagirov, Ann. Inst. Henri Poincaré **A 9** (1968) 109.

[4] A. Accioly and H. Blas, Phys. Rev. **D 66** (2002) 067501; Mod. Phys. Lett. **A 18** (2003) 867.

[5] J. P. Costella and B. H. J. McKellar, Am. J. Phys. **63** (1995) 1119; A. J. Silenko and O. V. Teryaev, Phys. Rev. **D 71** (2005) 064016; A. J. Silenko, Pis'ma Zh.

Fiz. Elem. Chast. Atom. Yadra **10** (2013) 144 [Phys. Part. Nucl. Lett. **10** (2013) 91].

[6] A. J. Silenko, J. Math. Phys. **44** (2003) 2952.

[7] A. J. Silenko, Phys. Rev. **A 77** (2008) 012116.

[8] A.J. Silenko, Teor. Mat. Fiz. **156** (2008) 398 [Theor. Math. Phys. **156** (2008) 1308].

[9] A. J. Silenko, Phys. Rev. **D 88** (2013) 045004.

[10] A. J. Silenko, *New symmetry properties of pointlike scalar and Dirac particles*, arXiv:1501.03635.

[11] A. J. Silenko, Teor. Mat. Fiz. **176** (2013) 189 [Theor. Math. Phys. **176** (2013) 987]; arXiv:1501.02052.

[12] Yu. N. Obukhov, A. J. Silenko, and O. V. Teryaev, Phys. Rev. **D 84** (2011) 024025.

Spontaneous supersymmetry breaking and instanton sum in 2D type IIA superstring theory*

Fumihiko Sugino[†]
Okayama Institute for Quantum Physics,
Kyoyama 1-9-1, Kita-ku, Okayama 700-0015, Japan

Abstract

We consider a double-well supersymmetric matrix model and its interpretation as a nonperturbative definition of two-dimensional type IIA superstring theory. The interpretation is confirmed by direct comparison of symmetries and amplitudes in both sides of the matrix model and the IIA superstring theory. Next, we obtain the full nonperturbative free energy of the matrix model in terms of the Tracy-Widom distribution in random matrix theory. Its weak coupling expansion implies spontaneous supersymmetry breaking due to instantons, and strong coupling behavior suggests the existence of a well-defined S-dual theory. Furthermore, from the expression of the free energy, we see a smooth connection between a non-supersymmetric string theory and the IIA superstring theory.

PACS codes: 02.10.Yn; 11.25.Pm; 11.25.Sq

1. Introduction

Solvable matrix models for two-dimensional quantum gravity or noncritical string theory had been vigorously investigated around 1990, focusing on nonperturbative aspects in string theory [5]. While this approach has been successful for bosonic string theory, little has been known for superstring theory, in particular theories possessing target-space supersymmetry (SUSY). So it would be worth while to considering (solvable) matrix models describing superstring theory with target-space SUSY. In this article, we discuss correspondence between a simple zero-dimensional SUSY double-well matrix model and two-dimensional type IIA superstring theory on a nontrivial Ramond-Ramond (RR) background. Then, nonperturbative effect of the matrix model is computed in its double scaling limit. As a result, we find that SUSY is spontaneously broken due to instantons in the matrix

* This is mainly based on the work with Tsunehide Kuroki, Michael G. Endres, Hiroshi Suzuki and Shinsuke M. Nishigaki [1, 2, 3, 4]. The work is supported in part by a Grant-in-Aid for Scientific Research (C), 25400289.
† e-mail address: fusugino@gmail.com

model. According to the correspondence, this suggests spontaneous SUSY breaking at the nonperturbative level in the type IIA superstring theory.

We hope our analysis is helpful to understand nonperturbative dynamics of matrix models of super Yang-Mills type for critical superstring theory [6, 7, 8].

2. Review of double-well SUSY matrix model

We give a brief review of a SUSY matrix model given by the following action [9]:

$$S = N\mathrm{tr}\left[\frac{1}{2}B^2 + iB(\phi^2 - \mu^2) + \bar{\psi}(\phi\psi + \psi\phi)\right], \tag{1}$$

where B and ϕ are $N \times N$ hermitian matrices, and ψ and $\bar{\psi}$ are $N \times N$ Grassmann-odd matrices. The action is invariant under SUSY transformations generated by Q and $\bar{Q}$:

$$Q\phi = \psi, \quad Q\psi = 0, \quad Q\bar{\psi} = -iB, \quad QB = 0, \tag{2}$$
$$\bar{Q}\phi = -\bar{\psi}, \quad \bar{Q}\bar{\psi} = 0, \quad \bar{Q}\psi = -iB, \quad \bar{Q}B = 0, \tag{3}$$

which leads to the nilpotency: $Q^2 = \bar{Q}^2 = \{Q, \bar{Q}\} = 0$. A large-$N$ saddle point for the eigenvalue distribution of the matrix ϕ: $\rho(x) \equiv \frac{1}{N}\mathrm{tr}\,\delta(x - \phi)$ has two supports $[-b, -a] \cup [a, b]$ in the case of $\mu^2 > 2$ (the two-cut solution) and a single support $[-c, c]$ for $\mu^2 < 2$ (the one-cut solution). Here, $a = \sqrt{\mu^2 - 2}$, $b = \sqrt{\mu^2 + 2}$ and $c = \sqrt{\frac{2}{3}}\left(\mu^2 + \sqrt{\mu^4 + 12}\right)^{1/2}$. The two-cut solution is characterized by the filling fractions (ν_+, ν_-) satisfying $\nu_+ + \nu_- = 1$, which indicate that the ratio of eigenvalues distributing over $[a, b]$ and those distributing over $[-b, -a]$ is $\nu_+ : \nu_-$. The large-N free energy and the expectation values $\left\langle \frac{1}{N}\mathrm{tr}\,B^n \right\rangle$ $(n = 1, 2, \cdots)$ evaluated at the two-cut solution turn out to all vanish [9], strongly suggesting that the solution preserves SUSY. Thus, we conclude that the SUSY minima are infinitely degenerate and parametrized by (ν_+, ν_-) in the simple large-N limit (the planar limit). On the other hand, the one-cut solution gives nonzero values of $\left\langle \frac{1}{N}\mathrm{tr}B \right\rangle$ and of the large-N free energy, showing that SUSY is broken [10]. The transition between the SUSY phase ($\mu^2 > 2$) and the SUSY broken phase ($\mu^2 < 2$) is of the third order. Namely, the third derivative of the free energy with respect to μ^2 has a jump at $\mu^2 = 2$.

In the SUSY phase ($\mu^2 > 2$) with the filling fraction (ν_+, ν_-), the planar correlation functions of the operators Φ_{2k+1} $(k = 0, 1, 2, \cdots)$ [1]:

$$\Phi_{2k+1} \equiv \frac{1}{N}\mathrm{tr}\,\phi^{2k+1} + (\text{mixing}) \tag{4}$$

[1] The second term "(mixing)" in (4) consists of lower even powers of ϕ. Its explicit form is given in ref. [1].

behave as [2]

$$\left\langle \prod_{i=1}^{n} \Phi_{2k_i+1} \right\rangle_{C,0} \sim (\nu_+ - \nu_-)^n \, u_{k_1,\cdots,k_n} \, \omega^{3+\sum_{i=1}^{n}(k_i-1)} \, (\ln\omega)^n \tag{5}$$

with $\omega \equiv \frac{1}{4}(\mu^2 - 2)$ and the constants $u_{k_1,\cdots,k_n}$ taking the form as

$$\begin{aligned} u_k &= \frac{2^{k+2}}{\pi} \frac{(2k+1)!!}{(k+2)!}, \\ u_{k,\ell} &= -\frac{1}{2\pi^2} \frac{1}{k+\ell+1} \frac{(2k+1)!}{(k!)^2} \frac{(2\ell+1)!}{(\ell!)^2}, \\ & \cdots . \end{aligned} \tag{6}$$

The planar two-point functions of fermions $\Psi_{2k+1} = \frac{1}{N}\mathrm{tr}\,\psi^{2k+1}$+(mixing) and $\bar{\Psi}^{2\ell+1} = \frac{1}{N}\mathrm{tr}\,\bar{\psi}^{2\ell+1}$ + (mixing) are evaluated as

$$\langle \Psi_{2k+1} \bar{\Psi}_{2\ell+1} \rangle_{C,0} \sim \delta_{k,\ell} \, v_k \, (\nu_+ - \nu_-)^{2k+1} \omega^{2k+1} \ln\omega \tag{7}$$

with v_k being constants $v_0 = \frac{1}{\pi}$, $v_1 = \frac{6}{\pi}$, $\cdots$.

3. 2D type IIA superstring

The type II superstring theory discussed in refs. [11, 12, 13] has the target space $(\varphi, x) \in$ (Liouville direction) $\times$ (S^1 with self-dual radius). The holomorphic energy-momentum tensor on the string world-sheet is

$$T = -\frac{1}{2}(\partial x)^2 - \frac{1}{2}\psi_x \partial \psi_x - \frac{1}{2}(\partial\varphi)^2 + \partial^2\varphi - \frac{1}{2}\psi_\ell \partial \psi_\ell \tag{8}$$

excluding ghosts' part. ψ_x and ψ_ℓ are superpartners of x and φ, respectively. Target-space supercurrents in the type IIA theory

$$q_+(z) = e^{-\frac{1}{2}\phi(z) - \frac{i}{2}H(z) - ix(z)}, \qquad \bar{q}_-(\bar{z}) = e^{-\frac{1}{2}\bar{\phi}(\bar{z}) + \frac{i}{2}\bar{H}(\bar{z}) + i\bar{x}(\bar{z})} \tag{9}$$

exist only on the S^1 target space of the self-dual radius. ϕ ($\bar{\phi}$) is the holomorphic (anti-holomorphic) bosonized superconformal ghost, and the fermions are bosonized as $\psi_\ell \pm i\psi_x = \sqrt{2}\, e^{\mp iH}$, $\bar{\psi}_\ell \pm i\bar{\psi}_x = \sqrt{2}\, e^{\mp i\bar{H}}$. In addition, we should care about cocycle factors in order to realize the anticommuting nature between q_+ and $\bar{q}_-$. Supercurrents with the cocycle factors are

$$\hat{q}_+(z) = e^{\pi\beta(\frac{1}{2}p_{\bar{\phi}} - i\frac{1}{2}p_{\bar{h}} - ip_{\bar{x}})}\, q_+(z), \qquad \hat{\bar{q}}_-(\bar{w}) = e^{-\pi\beta(\frac{1}{2}p_\phi + i\frac{1}{2}p_h + ip_x)}\, \bar{q}_-(\bar{w}), \tag{10}$$

[2]The suffix "C" means that connected parts are taken, and the symbol "$\sim$" denotes equality up to additive less singular terms.

where $\beta \in \mathbf{Z}+\frac{1}{2}$, and p_ϕ, p_h and p_x ($p_{\bar\phi}$, $p_{\bar h}$ and $p_{\bar x}$) are momentum modes of holomorphic part (anti-holomorphic part) of free bosons [2]. Then the supercharges

$$\hat{Q}_+ = \oint \frac{dz}{2\pi i}\,\hat{q}_+(z), \qquad \hat{\bar{Q}}_- = \oint \frac{d\bar{z}}{2\pi i}\,\hat{\bar{q}}_-(\bar{z}) \tag{11}$$

are nilpotent $\hat{Q}_+^2 = \hat{\bar{Q}}_-^2 = \{\hat{Q}_+, \hat{\bar{Q}}_-\} = 0$, which matches the property of the supercharges Q and $\bar{Q}$ in the matrix model.

The spectrum except special massive states is represented by the NS "tachyon" [3] vertex operator (in (-1) picture):

$$T_k = e^{-\phi+ikx+p_\ell\varphi}, \qquad \bar{T}_{\bar{k}} = e^{-\bar\phi+i\bar{k}\bar{x}+p_\ell\bar\varphi}, \tag{12}$$

and by the R vertex operator (in $(-\frac{1}{2})$ picture):

$$V_{k,\,\epsilon} = e^{-\frac{1}{2}\phi+\frac{i}{2}\epsilon H+ikx+p_\ell\varphi}, \qquad \bar{V}_{\bar{k},\,\bar\epsilon} = e^{-\frac{1}{2}\bar\phi+\frac{i}{2}\bar\epsilon\bar{H}+i\bar{k}\bar{x}+p_\ell\bar\varphi} \tag{13}$$

with $\epsilon, \bar\epsilon = \pm 1$. Cocycle factors for vertex operators are introduced as [2]

$$\begin{aligned}
&\hat{T}_k(z) = e^{\pi\beta(p_{\bar\phi}+ikp_{\bar x})}\,T_k(z), \quad \hat{\bar{T}}_{\bar k}(\bar z) = e^{-\pi\beta(p_\phi+i\bar{k}p_x)}\,\bar{T}_{\bar k}(\bar z),\\
&\hat{V}_{k,\,\epsilon}(z) = e^{\pi\beta(\frac{1}{2}p_{\bar\phi}+i\frac{\epsilon}{2}p_{\bar h}+ikp_{\bar x})}\,V_{k,\,\epsilon}(z),\\
&\hat{\bar{V}}_{\bar k,\,\bar\epsilon}(\bar z) = e^{-\pi\beta(\frac{1}{2}p_\phi+i\frac{\bar\epsilon}{2}p_h+i\bar{k}p_x)}\,\bar{V}_{\bar k,\,\bar\epsilon}(\bar z).
\end{aligned} \tag{14}$$

Locality with the supercurrents, mutual locality, superconformal invariance and the level matching condition determine physical vertex operators. As discussed in [13], there are two consistent sets of physical vertex operators - "momentum background" and "winding background". Here, we consider the "winding background". [4] The physical spectrum in the "winding background" is given by

$$\begin{aligned}
&(\text{NS, NS}): && \hat{T}_k\,\hat{\bar{T}}_{-k} && (k \in \mathbf{Z}+\tfrac{1}{2}),\\
&(\text{R+, R}-): && \hat{V}_{k,+1}\,\hat{\bar{V}}_{-k,-1} && (k = \tfrac{1}{2}, \tfrac{3}{2}, \cdots),\\
&(\text{R}-\text{, R+}): && \hat{V}_{-k,-1}\,\hat{\bar{V}}_{k,+1} && (k = 0, 1, 2, \cdots),\\
&(\text{NS, R}-): && \hat{T}_{-k}\,\hat{\bar{V}}_{-k,-1} && (k = \tfrac{1}{2}, \tfrac{3}{2}, \cdots),\\
&(\text{R+, NS}): && \hat{V}_{k,+1}\,\hat{\bar{T}}_k && (k = \tfrac{1}{2}, \tfrac{3}{2}, \cdots),
\end{aligned} \tag{15}$$

[3] In two dimensions, "tachyon" is not truely tachyonic but massless.

[4] We can repeat the parallel argument for "momentum background" in the type IIB theory, which is equivalent to the "winding background" in the type IIA theory through T-duality with respect to the S^1 direction.

where we take a branch of $p_\ell = 1 - |k|$ satisfying the locality bound $p_\ell \leq Q/2 = 1$ [14].

Comparing the SUSY transformation property of the matrix-model operators and that of the above vertex operators leads to the correspondence [1]:

$$\begin{aligned}
\Phi_{2k+1} &\Leftrightarrow \mathcal{V}_\phi(k) \equiv \int d^2z\, \hat{V}_{k+\frac{1}{2},+1}(z)\, \hat{\bar{V}}_{-k-\frac{1}{2},-1}(\bar{z}), \\
\Psi_{2k+1} &\Leftrightarrow \mathcal{V}_\psi(k) \equiv \int d^2z\, \hat{T}_{-k-\frac{1}{2}}(z)\, \hat{\bar{V}}_{-k-\frac{1}{2},-1}(\bar{z}), \\
\bar{\Psi}_{2k+1} &\Leftrightarrow \mathcal{V}_{\bar{\psi}}(k) \equiv \int d^2z\, \hat{V}_{k+\frac{1}{2},+1}(z)\, \hat{\bar{T}}_{k+\frac{1}{2}}(\bar{z}), \\
\frac{1}{N}\mathrm{tr}\,(-iB) &\Leftrightarrow \mathcal{V}_B(0) \equiv \int d^2z\, \hat{T}_{-\frac{1}{2}}(z)\, \hat{\bar{T}}_{\frac{1}{2}}(\bar{z})
\end{aligned} \tag{16}$$

for $k = 0, 1, 2, \cdots$. Note that (R−, R+) operators are singlets under the target-space SUSYs $\hat{Q}_+, \hat{\bar{Q}}_-$, and appear to have no counterpart in the matrix model side. Since the expectation value of operators measuring an RR charge $\langle \Phi_{2k+1} \rangle_0$ does not vanish as seen in (5), we conjecture that the matrix model corresponds to the type IIA theory on a nontrivial background of the (R−, R+) fields. As a check of the conjecture, introducing the (R−, R+) background in the form of vertex operators

$$\begin{aligned}
W_{\mathrm{RR}} &= (\nu_+ - \nu_-) \sum_{k\in\mathbf{Z}} a_k\, \mu_1^{k+1} \mathcal{V}_k^{\mathrm{RR}}, \\
\mathcal{V}_k^{\mathrm{RR}} &\equiv \begin{cases} \int d^2z\, \hat{V}_{k,-1}(z) \hat{\bar{V}}_{-k,+1}(\bar{z}) & (p_\ell = 1 - |k|,\ k \leq 0) \\ \int d^2z\, \hat{V}^{(\mathrm{nonlocal})}_{-k,-1}(z) \hat{\bar{V}}^{(\mathrm{nonlocal})}_{k,+1}(\bar{z}) & (p_\ell = 1 + |k|,\ k \geq 1) \end{cases}
\end{aligned} \tag{17}$$

with a_k being numerical constants, we consider correlation functions among integrated vertex operators $\mathcal{V}_i$ in the IIA theory on a nontrivial (R−, R+) background in the form:

$$\left\langle\!\!\left\langle \prod_i \mathcal{V}_i \right\rangle\!\!\right\rangle \equiv \left\langle \left(\prod_i \mathcal{V}_i \right) e^{W_{\mathrm{RR}}} \right\rangle. \tag{18}$$

When the strength of the background $(\nu_+ - \nu_-)$ is small, we can compute the correlation functions by expanding the background as $e^{W_{\mathrm{RR}}} = 1 + W_{\mathrm{RR}} + \frac{1}{2!}(W_{\mathrm{RR}})^2 + \cdots$. Results of the computation support the equivalence of the matrix model to the IIA superstring theory on the (R−, R+) background [2]. Interestingly, in the calculation of the IIA amplitudes, we see that the multiple powers of $\ln \omega$ in the matrix-model correlators (5) are identified with resonances among the external particles and the (R−, R+) background.

4. Nonperturbative SUSY breaking in the matrix model

In this section, we obtain the full nonperturbative free energy of the matrix model as the Tracy-Widom distribution in random matrix theory in the double scaling limit

$$N \to \infty, \quad \omega \to 0 \quad \text{with } s \equiv 4N^{2/3}\omega \text{ fixed.} \tag{19}$$

In its weakly coupled region (s: large), instanton effects can be seen in the matrix model which are nonperturbative in $1/N$. Although such effects are of the order e^{-N} and vanish in the simple large N limit, we will see that they are nonvanishing in the double scaling limit (19).

The partition function of the matrix model given by the action (1) is expressed as

$$\begin{aligned} Z &= \int d^{N^2}\phi\, e^{-N\frac{1}{2}\mathrm{tr}(\phi^2-\mu^2)^2}\, \det(\phi\otimes\mathbf{1}+\mathbf{1}\otimes\phi) \\ &= \tilde{C}_N \int \Big(\prod_{i=1}^N d\lambda_i\Big)\, \triangle(\lambda)^2 \prod_{i,j=1}^N (\lambda_i+\lambda_j)\, e^{-N\sum_{i=1}^N \frac{1}{2}(\lambda_i^2-\mu^2)^2}, \end{aligned} \tag{20}$$

after integrating out matrices other than ϕ. Here, $\mathbf{1}$ is an $N\times N$ unit matrix, λ_i $(i=1,\cdots,N)$ are eigenvalues of ϕ, and $\triangle(\lambda)$ denotes the Vandermonde determinant $\triangle(\lambda)=\prod_{i>j}(\lambda_i-\lambda_j)$. $\tilde{C}_N$ is an numerical factor depending only on N given by

$$\frac{1}{\tilde{C}_N} = \int \Big(\prod_{i=1}^N d\lambda_i\Big)\, \triangle(\lambda)^2\, e^{-N\sum_{i=1}^N \frac{1}{2}\lambda_i^2} = (2\pi)^{\frac{N}{2}} \frac{\prod_{k=0}^N k!}{N^{\frac{N^2}{2}}}. \tag{21}$$

Contributions to the partition function are divided by sectors labeled by the filling fraction (ν_+,ν_-) as

$$Z = \sum_{\nu_- N=0}^{N} \frac{N!}{(\nu_+ N)!(\nu_- N)!} Z_{(\nu_+,\nu_-)} \tag{22}$$

with

$$\begin{aligned} Z_{(\nu_+,\nu_-)} &\equiv \tilde{C}_N \int_0^\infty \Big(\prod_{i=1}^{\nu_+ N} d\lambda_i\Big) \int_{-\infty}^0 \Big(\prod_{j=\nu_+ N+1}^{N} d\lambda_j\Big) \Big(\prod_{n=1}^N 2\lambda_n\Big) \\ &\quad \times \Big\{\prod_{n>m} (\lambda_n^2-\lambda_m^2)^2\Big\}\, e^{-N\sum_{i=1}^N \frac{1}{2}(\lambda_i^2-\mu^2)^2}. \end{aligned} \tag{23}$$

Here, it is easy to see

$$Z_{(\nu_+,\nu_-)} = (-1)^{\nu_- N} Z_{(1,0)}, \tag{24}$$

which leads to the vanishing partition function:

$$Z = (1 + (-1))^N Z_{(1,0)} = 0. \tag{25}$$

In order for expectation values normalized by the partition function to be well-defined, we regularize the partition function by introducing a factor $e^{-i\alpha\nu_- N}$ with small α in front of $Z_{(\nu_+,\nu_-)}$. The regularized partition function reads

$$Z_\alpha \equiv \sum_{\nu_- N=0}^{N} \frac{N!}{(\nu_+ N)!(\nu_- N)!} e^{-i\alpha\nu_- N} Z_{(\nu_+,\nu_-)} = (1 - e^{-i\alpha})^N Z_{(1,0)}. \tag{26}$$

Notice that calculations in perturbation theory of $1/N$ in section 2. concern the partition function in a single sector $(Z_{(\nu_+,\nu_-)})$, in which such a regularization was not needed. On the other hand, since nonperturbative contributions to be computed here possibly communicate among various sectors of filling fractions, we should consider the total partition function (22) and its null result requires the regularization.

The expectation value of $\frac{1}{N}\mathrm{tr}(iB)$ under the regularization (26) is expressed as

$$\left\langle \frac{1}{N}\mathrm{tr}\,(iB) \right\rangle_\alpha = \frac{1}{N^2}\frac{1}{Z_\alpha}\frac{\partial}{\partial(\mu^2)}Z_\alpha = \frac{1}{N^2}\frac{1}{Z_{(1,0)}}\frac{\partial}{\partial(\mu^2)}Z_{(1,0)} \tag{27}$$

due to a cancellation of the factor $(1-e^{-i\alpha})^N$ in (26) between the numerator and the denominator. The regularized expectation value $\left\langle \frac{1}{N}\mathrm{tr}\,(iB) \right\rangle_\alpha$ is independent of α and well-defined in the limit $\alpha \to 0$, and thus serves as an order parameter for spontaneous SUSY breaking.

4.1. Tracy-Widom distribution

Under the change of variables $x_i = -\lambda_i^2 + \mu^2$, the partition function $Z_{(1,0)}$ defined in (23) reduces to Gaussian matrix integrals

$$Z_{(1,0)} = \tilde{C}_N \int_{-\infty}^{\mu^2} \left(\prod_{i=1}^{N} dx_i \right) \triangle(x)^2 \, e^{-N\sum_{i=1}^{N}\frac{1}{2}x_i^2}. \tag{28}$$

It seems almost trivial, but a nontrivial effect arises from the upper bound of the integration region. Techniques in random matrix theory [15] give a closed form for the partition function in the double scaling limit (19):

$$F(s) = -\ln Z_{(1,0)} = \int_s^\infty (x-s)q(x)^2 dx, \tag{29}$$

where $q(x)$ satisfies a Painlevé II differential equation

$$q''(x) = xq(x) + 2q(x)^3 \tag{30}$$

with the boundary condition

$$q(x) \to \mathrm{Ai}(x) \qquad (x \to +\infty). \tag{31}$$

Such a solution is unique and known as the Hastings-McLeod solution [16]. Since eq. (19) indicates that the string coupling constant $g_s \sim 1/N$ is proportional to $s^{-3/2}$, the region of $s \gg 1$ $(0 < s \ll 1)$ describes the weakly (strongly) coupled IIA strings.

4.2. Weak coupling expansion

The partition function is given by the Fredholm determinant of the Airy kernel [15]

$$Z_{(1,0)} = \mathrm{Det}(1 - \hat{K}_{\mathrm{Ai}}|_{[s,\infty)}), \tag{32}$$

where the operator $\hat{K}_{\mathrm{Ai}}|_{[s,\infty)}$ can be represented as the integration kernel on the interval $[s, \infty)$:

$$K_{\mathrm{Ai}}(x,y) \equiv \frac{\mathrm{Ai}(x)\mathrm{Ai}'(y) - \mathrm{Ai}'(x)\mathrm{Ai}(y)}{x-y}. \tag{33}$$

From the above fact, it turns out that the weak coupling expansion (large-s expansion) of the free energy is expressed as an instanton sum [4]

$$F = -\ln Z_{(1,0)} = \sum_{k=1}^{\infty} F_{k-\mathrm{inst.}} \tag{34}$$

with

$$\begin{aligned} F_{k-\mathrm{inst.}} &= \frac{1}{k}\int_s^\infty dt_1 \ldots dt_k\, K_{\mathrm{Ai}}(t_1,t_2)\, K_{\mathrm{Ai}}(t_2,t_3)\cdots K_{\mathrm{Ai}}(t_k,t_1) \\ &\sim \frac{1}{k}\left(\frac{1}{16\pi s^{3/2}}\, e^{-\frac{4}{3}s^{3/2}}\right)^k \left[1 + a_1^{(k)} s^{-3/2} + a_2^{(k)} s^{-3} + \cdots\right]. \end{aligned} \tag{35}$$

Some of the coefficients are

$$\begin{aligned} a_1^{(1)} &= -\frac{35}{24}, & a_2^{(1)} &= \frac{3745}{1152}, & a_3^{(1)} &= -\frac{805805}{82944}, \cdots \\ a_1^{(2)} &= -\frac{35}{12}, & a_2^{(2)} &= \frac{619}{72}, & a_3^{(2)} &= -\frac{592117}{20736}, \cdots \\ a_1^{(3)} &= -\frac{35}{8}, & a_2^{(3)} &= \frac{2059}{128}, & a_3^{(3)} &= -\frac{184591}{3072}, \cdots \\ a_1^{(4)} &= -\frac{35}{6}, & a_2^{(4)} &= \frac{3701}{144}, & a_3^{(4)} &= -\frac{1112077}{10368}, \cdots. \end{aligned} \tag{36}$$

The contribution to the free energy has no perturbative part and starts from nonperturbative effects of the instanton action $\frac{4}{3}s^{3/2} \propto N$ and its fluctuations expanded by $s^{-3/2} \propto N^{-1}$. It seems plausible that the non-perturbative contributions are provided by D-brane like objects. The order parameter of the SUSY breaking (with the wave function renormalization factor $N^{4/3}$), $N^{4/3}\left\langle \frac{1}{N}\mathrm{tr}(iB)\right\rangle^{(1,0)} = -F'(s)$, remains nonzero, implying that the target-space SUSY in the two-dimensional IIA theory is spontaneously broken by D-brane like objects. Corresponding Nambu-Goldstone fermions are identified with $\frac{1}{N}\mathrm{tr}\bar{\psi}$ and $\frac{1}{N}\mathrm{tr}\psi$ associated with the breaking of Q and $\bar{Q}$, respectively [3].

4.3. Strong coupling expansion

The Taylor series expansion of (29) around $s = 0$ is

$$F(s) = 0.0311059853 - 0.0690913807s + 0.0673670913s^2 - 0.0361399144s^3 + \cdots, \tag{37}$$

which gives strong coupling expansion of the IIA superstring theory. The strong coupling limit is regular and finite. In particular, the expression is smooth around $s = 0$ and there is no obstruction to be continued to the $s < 0$ region (i.e. $\mu^2 < 2$), whereas in section 2. we had seen the third order phase transition across the point $\mu^2 = 2$ in the planar limit. Thus, the singularity in the planar limit becomes smeared out in the double scaling limit. [5] In the string-theory perspective, singular behavior at the string tree level is smoothed out by quantum effects. Similar phenomenon can be seen in the unitary one-matrix model [17].

In the region of $s < 0$, the free energy has a perturbative series in $(-s)^{-3} \propto N^{-2}$, which seems to allow an interpretation as non-SUSY (type 0) string theory. Thus, we can see that the matrix model describes both of the IIA superstrings and type 0 strings in a unified manner at least concerning the free energy.

5. Summary and Discussion

We have computed planar correlation functions in the double-well SUSY matrix model, and discussed its correspondence to two-dimensional type IIA superstring theory on (R−,R+) background. This is an interesting example of matrix models for superstrings with target-space SUSY, which allows explicit calculation of various amplitudes of observables not restricted to those protected by SUSY.

It is interesting to examine the correspondence at deeper level, for instance in higher genus or higher point amplitudes and in amplitudes containing special massive operators. It is also important to discuss the correspondence in the off-shell formulation such as the hybrid formalism [18].

[5]From a physical viewpoint, it is considered that the free energy in the $\nu_+ = \nu_- = 1/2$ sector for $\mu^2 > 2$ smoothly connects to the free energy for $\mu^2 < 2$ in the double scaling limit. Due to (24), the former is essentially equal to (29) except an unimportant additive term.

Next, we have explicitly presented the full nonperturbative expression of the matrix-model free energy in a closed form. The result in the weakly coupled regime shows that the SUSY is spontaneously broken by nonperturbative effects due to instantons. In particular, the instanton effects survive in the double scaling limit, which implies that SUSY breaking takes place by nonperturbative dynamics in the target space of the type IIA superstring theory. It is interesting to reproduce the instanton contributions from dynamics of D-branes in the type IIA theory.

The free energy in the strongly coupled limit is smooth, which means that the singular behavior of the third order phase transition in the planar limit (at the string tree level) becomes smeared out in the double scaling limit (including quantum effects in the IIA strings). The regularity at $s = 0$ might suggest the existence of an S-dual theory, and the region $s < 0$ seems to describe a non-SUSY (type 0) string theory. It would be intriguing to identify such S-dual theory and to understand the moduli space of noncritical string theories given by the matrix model.

Acknowledgments

The author would like to thank the organizers of the 8th Mathematical Physics Meeting, especially Professors Branko Dragovich, Zoran Rakic and Igor Salom, for the invitation to the wonderful meeting and for warm hospitality.

References

[1] T. Kuroki and F. Sugino, *Nucl. Phys.* **B 867** (2013) 448.

[2] T. Kuroki and F. Sugino, *J. High Energy Phys.* **1403** (2014) 006.

[3] M. G. Endres, T. Kuroki, F. Sugino and H. Suzuki, *Nucl. Phys.* **B 876** (2013) 758.

[4] S. M. Nishigaki and F. Sugino, *J. High Energy Phys.* **1409** (2014) 104.

[5] For a review, see P. Di Francesco, P. H. Ginsparg and J. Zinn-Justin, *Phys. Rept.* **254** (1995) 1.

[6] T. Banks, W. Fischler, S. H. Shenker and L. Susskind, *Phys. Rev.* **D 55** (1997) 5112.

[7] N. Ishibashi, H. Kawai, Y. Kitazawa and A. Tsuchiya, *Nucl. Phys.* **B 498** (1997) 467.

[8] R. Dijkgraaf, E. P. Verlinde and H. L. Verlinde, *Nucl. Phys.* **B 500** (1997) 43.

[9] T. Kuroki and F. Sugino, *Nucl. Phys.* **B 830** (2010) 434.

[10] T. Kuroki and F. Sugino, *Nucl. Phys.* **B 844** (2011) 409.

[11] D. Kutasov and N. Seiberg, *Phys. Lett.* **B 251** (1990) 67.

[12] S. Murthy, *J. High Energy Phys.* **0311** (2003) 056.

[13] H. Ita, H. Nieder and Y. Oz, *J. High Energy Phys.* **0506** (2005) 055.

[14] N. Seiberg, *Prog. Theor. Phys. Suppl.* **102** (1990) 319.

[15] C. A. Tracy and H. Widom, *Commun. Math. Phys.* **159** (1984) 151.

[16] S. P. Hastings and J. B. McLeod, *Arch. Rat. Mech. Anal.* **73** (1980) 31.

[17] V. Periwal and D. Shevitz, *Phys. Rev. Lett.* **64** (1990) 1326.

[18] P. A. Grassi and Y. Oz, *Preprint* arXiv:hep-th/0507168

Dynamical Emergence of FRW Cosmological Models

Marek Szydłowski*
Astronomical Observatory, Jagiellonian University,
Orla 171, 30-244 Kraków, Poland
Mark Kac Complex Systems Research Centre, Jagiellonian University,
Reymonta 4, 30-059 Kraków, Poland

Paweł Tambor†
Copernicus Center for Interdisciplinary Studies,
Gronostajowa 3, 30-387 Kraków, Poland

Abstract

Recent astronomical observations strongly indicate that the current Universe is undergoing an accelerated phase of expansion. The discovery of this fact was unexpected and resulted in the comeback of cosmological constant. The conception of standard cosmological model has its roots in this context. The paper relates to the methodological status of effective theories in the context of cosmological investigations. We argue that the standard cosmological model (LCDM model) as well as the CDM have a status of effective theories only, similarly to the standard model of particle physics. The LCDM model is studied from the point of view of the methodological debate on reductionism and epistemological emergence in the science. It is shown in the paper that bifurcation as well as structural instability notion can be useful in the detection of emergence the LCDM model from the CDM model. We demonstrate that the structural stability of the LCDM model can explain the flexibility of the model to accommodation of the observational data. We show that LCDM model can be derived from CDM as the bifurcation. It is an example of acausal derivation of Lambda term. The case study of emergence of LCDM model suggests that it can be understood in terms of bifurcation and structural stability issue. The reduction from the upper models represented in terms of dynamical system to low-level ones can be realized in any case by application of a mathematical limit (boundary crossing) with respect to the model parameter. It is a simple consequence of mathematical theorem about smooth dependence solutions with respect to time, initial condition and the parameters.

Keywords: cosmological dynamical systems, emergence, standard cosmological model.

* e-mail address: marek.szydlowski@uj.edu.pl
† e-mail address: pawel.tambor@gmail.com

1. Introduction

Recent astronomical observations of distant type SNIa supernovae strongly indicate that the current Universe is undergoing an accelerated phase of expansion [1, 2]. If the Universe's evolution is described by homogeneous and isotropic models filled with a perfect fluid, then the acceleration should be driven by a perfect fluid violating the strong energy condition. If different candidates for a fluid termed dark energy are suggested, the simple candidate for the dark energy in the form of positive cosmological constant seems to be the best one [19]. While the Lambda CDM model is a suitable to phenomenologically describe the acceleration phase of the expansion of the Universe there is a serious problem with the interpretation of the Lambda term as a quantum vacuum energy due to the fine tuning problem. Our studies show that when the LCDM has the status of an effective theory, which offers description of the observational facts rather than their explanation, it introduces principally a new theoretical element which plays the role of an effective parameter changing the dynamics dramatically.

Our main result is that structural stability notion taken from the dynamical system theory may be useful in our understanding of the emergence of CDM to LCDM model as well as in understanding the reduction of LCDM to CDM one. We argue that the concepts of structural stability might be a suitable setup for the methodology of cosmology discussion and that the notion of bifurcation successfully replaces the notion of emergence.

The LCDM model should in our opinion be treated as an effective theory for the following reasons. Firstly, theory of gravity which describes the gravitational sector of cosmology is very complicated, but if we postulate some simplified assumption like symmetry assumption idealization, then we obtain the simplest model which can be represented in the form of the dynamical system. In the cosmology, assumption of homogeneity and isotropy of space like sections of constant cosmic time ($t = const$) seems to be justified by the distribution of large scale structure of astronomical objects (cosmological principle). If we assume that universe is homogeneus and isotropic the evolution of it is determined by a single function of $a(t)$ called the scale factor. If we postulate that source of gravity is in the form of perfect fluid with energy density $\varrho(t)$ and pressure $p(t)$, then Einstein field equation reduces to the ordinary system of differential equation determining a single function $a(t)$. These equations called Friedmann equations FRW model describe the evolution of the Universe at the large scale. Note that this is impossible basing on general relativity without the symmetry assumption where there is universal time conception. Finally, the general acceptance of the LCDM model as working is a good strategy [4] but one may also seek alternative physics (pragmatism).

2. Structural stability issues

If the space satisfies cosmological principle then Einstein field equations can be reduced to the system of ordinary differential equations, i.e. dynamical systems. Hence the dynamical system methods to the cosmology could be applied to cosmology in a natural way. The application of these methods allows us to reveal some stability properties of particular model visualized in geometrical way as the trajectories in the phase space. Therefore one can see how large the class of solutions leading to the desired property in tools of the attractors and the inset of limit set is (an attractor is a limit set with an open inset – all the initial conditions that end up in the some equilibrium state). The attractors are the most prominent experimentally. It is because the probability of an initial state of the experiment to evolve asymptotically to the limit set is proportional to the volume of inset.

The idea, now called structural stability, emerged early in the history of dynamics investigation in 1930s the writings of Andronov, Leontovich and Pontryagin in Russia (1934) [5] (the authors do not use the name *structural stability*, but rather the name "roughly systems"). This idea is based on an observation of an actual state of the system which can never be specified exactly and application of the dynamical systems might be useful anyway if it can describe the features of the phase portrait that persist when the state of the system is allowed to move around.

Among all dynamicists there are shared prejudices that: (1) there is a class of phase portraits that are far simpler than arbitrary ones. It explains why a considerable portion of the mathematical physics has been dominated by the search for the generic properties. The exceptional case should not arise very often in application and they de facto interrupt discussion (classification); (2) The physically realistic models of the world should possess some kind of the structural stability because having so many dramatically different models all agreeing with observation would be fatal for the empirical method of science (see also [6, 7].

These prejudices can in the Holton's terminology be treated as thematic principles [8, 9]. In the cosmology a property (for example acceleration) is believed to be "physically realistic" if it can be attributed by the generic subsets of the models within a space of all admissible solutions or if it possesses a certain stability, i.e. if it is shared by a "epsilon perturbed model".

The dynamical system is called structurally stable if all δ-perturbation of it (sufficiently small) have the epsilon equivalent phase portrait. Therefore for the conception of structural stability we considered a δ-perturbation of vector field determined by the right-hand sides of the system which is small (measured by delta). We also need a conception of the epsilon equivalence. It takes the form of topological equivalence–a homeomorphism of the state space preserving the arrow of time on each trajectory. In the definition, the structural stability considers only the deformation of "rubber sheet" type stretches or slides the phase space a small amount measured by epsilon. The main advantage of the structural stability is that it is the characterization of global dynamics itself.

Recently properties of structural stability of cosmological models were investigated by S. Kokarev [10]. In the introduction to the paper, the author claims that the history of cosmology shows that corrections of cosmological models are realized mainly by the sequence of their, in a sense "small", modifications and some of them may "survive" after small changes, while others may disappear. In the former case the property is referred to as "rough" or structurally stable, in the later one – "thin" or structurally unstable. The author studies how some model properties, like singularities for example, will be present in the model if we "perturb" the model (e.g. generalize the Lagrangian of General Relativity). In our approach the property of structural stability is the property of the model itself. Also the type of perturbations is not specified (epsilon perturbation idea). Therefore, if we prove the structural instability of CDM model, the result will not depend on the choice of the type of perturbation. Then the property of structural stability becomes its constitutive property at the very beginning without restriction to the class of perturbation induced by considering new theories with generalized Lagrangian.

The idea of structural stability attempts to define the stability notion of differential deterministic models of the physical processes. In the case of planar dynamical systems (as in the case of models under consideration) there Peixoto theorem is true (Peixoto 1982) [11] which states that structurally stable dynamical systems form open and dense subsets in the space of all dynamical systems defined on the compact manifold. This theorem is a basic characterization of structurally stable dynamical systems in the plane which offers the possibility of the exact definition of generic (typical) and nongeneric (exceptional) cases (properties) by means of the notion of structural stability. Unfortunately there is no counterpart of this theorem in a more dimensional case when structurally unstable systems can also form open and dense subsets. For our aims, it is important that Peixoto theorem can give the characterization of generic cosmological models in terms of potential function V of the scale factor a which determine the motion of the system of Newtonian type: $\ddot{a} = -\frac{\partial V}{\partial a}$.

Therefore we can treat FRW equation with various forms of dark energy as the two-dynamical system which looks like Newtonian type where the role of coordinate variable is played by the cosmological radius (or redshift z: $1 + z = \frac{a_0}{a} \equiv x^{-1}$). We can construct an effective potential (the second order acceleration equation has exactly the Newtonian form) where the role of a coordinate variable is played by the cosmological radius.

Using the term of the structural stability first introduced by Andronov, Leontovich and Pontryagin in 1930s, one can classify different models of cosmic acceleration. It will be demonstrated that models with the accelerating phase which follow the deceleration are natural and typical from the point of view of the dynamical systems theory combined with the notion of structural stability in contrast to the models with bounces.

Let us introduce the following definition:

Definition 1 *If the set of all vector fields* $\boldsymbol{f} \in C^r(\mathcal{M})$ $(r \geq 1)$ *having a certain property contains an open dense subset of* $C^r(\mathcal{M})$, *then the property is called generic.*

From the physical point of view it is interesting to know whether certain subset v of $C^r(\mathcal{M})$ (representing a class of cosmological accelerating models in our case) contains a dense subset because it means that this property (acceleration) is typical in V.

It is not difficult to establish some simple relation between the geometry of potential function and the localization of critical points and its character for the case of dynamical systems of Newtonian type:

1. The critical point of the system under consideration $\dot{x} = y$, $y = -\frac{\partial V}{\partial x}$ lies always on x-axis, i.e. they represent static universe $y_0 = 0$, $x = x_0$;
2. The point $(x_0, 0)$ is a critical point of the Newtonian system if it is a critical point of the potential function $V(x)$, i.e. $V(x) = E$ (E is total energy of the system $E = \frac{y^2}{2} + V(x)$; $E = 0$ for the case flat models and $E = \frac{-k}{2}$ in general);
3. If $(x_0, 0)$ is a strict local maximum of $V(x)$, it is a saddle type critical point;
4. If $(x_0, 0)$ is a strict local minimum of the analytic function $V(x)$, it is the center;
5. If $(x_0, 0)$ is a horizontal inflection point of the $V(x)$, it is a cusp.

Therefore the geometry of potential function determines the critical points as well as its stability. The integral of energy defines the algebraic curves in the phase space (x, y) which are representing the evolution of the system with time. In any case the eigenvalues of the linearization matrix satisfy the characteristic equation $\lambda^2 + \frac{\partial^2 V}{\partial x^2}|_{x=x_0} = 0$.

3. Cosmological models as dynamical systems

Cosmology is based on the Einstein field equation which represents a very complicated system of partial nonlinear differential equations. Fortunately, the majority of main class of cosmological models from the point of view of observational data, belong to the class of the spatially homogeneous ones, for which the absolute cosmological time makes sense. As a consequence, the evolution of such models can be reduced to the systems of ordinary differential equations. Hence the methods of dynamical system theory or qualitative theory of differential equation can be naturally applied to cosmology. Among these classes of models especially interesting are the cosmological models with maximally symmetric space sections, i.e. homogeneous and isotropic. They are called FRW models (Friedmann-Robertson-Walker) if source of the gravity is a perfect fluid described in terms of energy density ϱ and pressure p, both are the functions of cosmological time t. The FRW dynamics is described by two basic equations:

$$\ddot{a} = -\frac{1}{6}(\varrho + 3p)a = -\frac{\partial V}{\partial a} \tag{1}$$

$$\dot{\varrho} = -3H(\varrho + p);, \tag{2}$$

where the potential $V = -\frac{1}{6}\varrho a^2$, a is the scale factor and $H = d\ln a/dt$ is Hubble's function, an overdot means the differentiation with respect to the cosmological time t.

The first equation is a consequence of the Einstein equations for the component (1,1), (2,2), (3,3) and the energy momentum tensor $T^{\mu}_{\nu} = \mathrm{diag}|-\varrho, p, p, p|$. This equation is called the Raychaudhuri or acceleration equation. The second equation represents the conservation condition

$T^{\mu}_{\nu;\mu} = 0$. It is very strange and unreasonable that such two simple equations satisfactorily describe the Universe's evolution at the large scales. Of course there is a more general class of cosmological models called the Bianchi models which have only the symmetry of homogeneity, but they fail to describe the current Universe which is isotropic as indicated measurement of the cosmic microwave background (CMB) radiation.

The system of equations (1) and (2) admits the first integral called the Friedmann equation

$$\varrho - 3H^2 = 3\frac{k}{a^2}, \tag{3}$$

where k is curvature constant $(0, \pm 1)$ and ϱ plays the role of effective energy density.

If we consider the Lambda CDM model, then

$$\varrho_{\text{eff}} = \varrho_{\text{m},0}a^{-3} + \Lambda, \tag{4}$$

i.e. energy density is the sum of dust matter (cold) and dark energy. Therefore the potential function for the flat FRW model assumes the following form:

$$V = -\frac{\varrho_{\text{eff}}a^2}{6} = (-)\{\varrho_{\text{m},0}a^{-1} + \Lambda a^2\}. \tag{5}$$

Formally the curvature effects as well as the cosmological constant term can be incorporated into the effective energy density ($\varrho_k = -\frac{k}{a^2}$; $\varrho_\Lambda = \Lambda$; $p_\Lambda = -\Lambda$, $p_k = -\frac{1}{3}\varrho_k$).

The form of equation (1) suggests the possible interpretation for the evolutional paths of cosmological models as a motion of a fictitious particle of unit mass in a one-dimensional potential parameterized by the scale factor. Following this interpretation the Universe is accelerating in the domain of configuration space $\{a\colon a \geq 0\}$ in which the potential is a decreasing function of the scale factor. In the opposite case if potential is a growing function of a the Universe is decelerating. The limit case of zero acceleration corresponds to an extremum of the potential function.

4. Emergence of the LCDM model from the CDM model in the framework of structural stability

It should be mentioned that the notion of emergence applied to physics shows quite different results than in philosophy. We suggest that while having philosophical flavor, the emergence notion should be treated in philosophy of science with great caution. It appears to be used in physics rather in a colloquial and informal way and emergence in the context of the cosmology, as we show in this paper, does not mean irreducibility or unpredictability.

The dynamical system investigation of the solutions of differential equations shifts its key point from founding and analyzing of individual solutions to investigating the space of all solutions for all admissible initial conditions, in the geometrical language of the phase space. Certain property (such as acceleration, singularities, etc.) is believed to be realistic if it can be attributed to a large subset of models within the space of all solutions [7]. The evolutional scenarios are represented by the phase curves or by critical points, limit circles or other limit sets. We say that two dynamical systems (or equivalently vector fields), say $f(x)$ and $g(x)$) are equivalent, if there is an orientation preserving homeomorphism sending integral curves of f into those of g. Of course this equivalence relation divided space of all dynamical systems on the plane on disjoint class of abstraction. Let phase space $E = R^n$, then ϵ - perturbation f is the function $g \in C^1(\mathcal{M})$ satisfying $\|f - g\|_1 < \epsilon$; where $\mathcal{M}$ is open subset of R^n and $\|...\|_1$ is C^1 norm form the Banach space. In the introduced language it is natural to formulate an idea of structural stability. The intuition is very simple, namely f is structurally stable vector field if for any vector field f and g are topologically equivalent. Then one can define the property of structural stability of the system.

Definition 2 *A vector field $f \in C^1(\mathcal{M})$ is called structurally stable if there is an $\epsilon > 0$ such that for all $g \in C^1(\mathcal{M})$ with $\|f - g\|_1 < \epsilon$, f and g are topologically equivalent on open subset R^n called $\mathcal{M}$.*

The 2-dimensional case is distinguished by the fact that the Peixoto theorem (1962) gave a complete characterization of structurally stable systems on any compact, two dimensional space asserts that they are generic, i.e. forms open and dense subsets in the space of all dynamical system on the plane [11].

While there is no counterpart to the Peixoto theorem in higher dimension, it can be easy used to test whether such dynamical systems or cosmological origin has a structurally stable global phase portrait. In particular, a vector field on the Poincaré sphere will be structurally unstable if there are non-hyperbolic critical points at infinity on the equator of the Poincaré sphere. In the opposite case if additionally the number of critical points and limit cycles is finite, f is structurally stable on S^2.

In the next section we prove that the CDM model is structurally unstable (therefore exceptional in the space of all dynamical systems on the plane) and its transition (which we called emergence) to the Lambda CDM model means such perturbation of the CDM system that new perturbed system is structurally stable (therefore generic).

5. Emergence of new properties of the model through a bifurcation

In our discussion it would be useful to consider a common approach to reduction in physics, so called *deductive criterion of reducibility* of Nagel [13]. In this concept reduction is a relation of derivation between upper-level and base-level theories. This is exactly our point of view from the case study of cosmology as based on the dynamical systems theory that the dynamical model from upper-level can be smoothly reduced to the

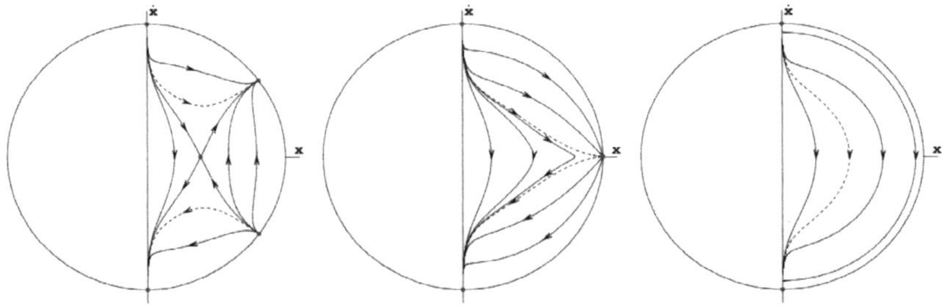

Figure 1: The phase portraits for different perturbations of the CDM model – from the left: 1) the LCDM model with the positive cosmological constant, 2) the LCDM model with the negative cosmological constant, 3) the CDM model with the vanishing cosmological constant. Note that only the systems with the cosmological constant are structurally stable, while the CDM model is unstable because of the presence of degenerated critical points at the circle at infinity (case 3). The right figure represents the Einstein static universe $(x, \dot{x}) = (\infty, 0)$. The critical point $(0, \infty)$ represents the big-bang singularity (an unstable node).

base-level one through the boundary crossing[1]. It is the example of a case where the emergence and reduction coexist.

Let us consider two models which must be connected using the cosmological parameter. This parameter plays the role of control parameter in the model and we assume that it assumes zero (vanishes) in the basal model. We are looking for weakly emergent properties of the model which can be derived (via bifurcation) from the complete knowledge of the basal model information. In order to do so we use bifurcation theory, from which information about new unveiling properties of the system can be predicted, at least in principle as we change the control parameter. Then in principle we can derive the system behavior because we can perform bifurcation analysis answering the question how the structure of the phase space qualitatively changes as parameter Λ is moved. Due to bifurcation we can not predict its future behavior with complete certainty (because there are to possible predictions concerning the value of cosmological constant; $\Lambda > 0$ or $\Lambda < 0$). Such a point of view seems to be very close to traditional conceptions of emergence (Broad, Popper, Nagel) that focus on unpredictability properties of upper-level model even given complete basal information. In the case of LCDM model novelty element is related with Λ-term, which emerges as a essential model parameter due to astronomical observation of type Ia distant supernovae (let us call it the empirical emergence).

Let us illustrate our point of view in the a very simple, yet instructive, example. The dynamics of the flat cosmological models with R-W sym-

[1]Which is granted by theorems from differential equations theory about smooth dependence solutions on the time, initial conditions and the parameters.

metry of space-like section, cosmological constant and without the matter (only for simplicity of presentation) is governed by a very simple equation (one-dimensional system)

$$\dot{x} = -x^2 + \frac{\Lambda}{3}, \tag{6}$$

where $x = H$ is Hubble parameter which measures the average rate of expansion of the Universe; Λ is here the cosmological constant parameter; $\dot{}$ denotes differentiation with respect to the cosmological time.

Obviously, the above system can be simply integrated in quadratures. Calculation gives

$$x(t) = \sqrt{\frac{\Lambda}{3}} \coth \sqrt{\frac{\Lambda}{3}}(t + C) \tag{7}$$

and $x(t = -C) = \infty$, where C is integration constant. Equation (6) can be also integrated for the special case of $\Lambda = 0$

$$x(t) = \frac{1}{t - t_0}. \tag{8}$$

Naive looking at the formula (7) gives rise to statement that there is no transition from the solution (7) to (8) as $\Lambda \to 0$. It is not a true point of view. The limit solution as $\Lambda \to 0$ can be achieved in the following way. Let us consider Λ in the formula (7) as small parameter and both functions cosh(...) i sinh(...) can be expanded in Taylor series with respect to $\sqrt{\frac{\Lambda}{3}}t$ term. Conserving linear parts of both functions, we obtain:

$$x(t) = \sqrt{\frac{\Lambda}{3}} \left[\frac{1 + \sqrt{\frac{\Lambda}{3}}(t + C) + 1 - \sqrt{\frac{\Lambda}{3}}(t + C)}{1 + \sqrt{\frac{\Lambda}{3}}(t + C) - 1 + \sqrt{\frac{\Lambda}{3}}(t + C)} \right] = \frac{1}{t + C}. \tag{9}$$

Now, if we put $C = -t_0$, then the solution for CDM is reproduced.

By this example one can observe how some small changes of the right hand side of the system dramatically change its solution. As a result in this system and solution emerge new asymptotic states representing de Sitter model.

The bifurcation theory serves to clarify the emergence of new properties (sometimes unexpected) of the system without solving this equation. Let us consider the system in the framework of bifurcation theory. For $\Lambda > 0$ there are two critical points $\dot{x} = 0$ at $x \pm \sqrt{\frac{\Lambda}{3}}$. From the physical point of view they are representing de Sitter model (expanding and contracting). Derivative $f(x)$ ($\dot{x} = f(x)$), $Df(x, \mu) = -2x$ and $Df(\pm\sqrt{\frac{\Lambda}{3}}, \Lambda) = \mp 2\sqrt{\frac{\Lambda}{3}}1$ and we can see that the critical point at $x = \sqrt{\frac{\Lambda}{3}}$ is stable while the critical point $x = -\sqrt{\frac{\Lambda}{3}}$ is unstable. For $\Lambda = 0$, there is only one critical point at

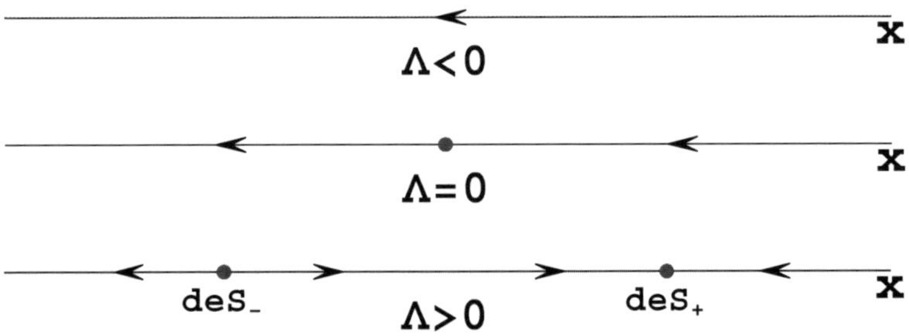

Figure 2: The phase portraits for flat FRW model without matter and with the cosmological constant of different signs. For $\Lambda < 0$ there is no critical point. For $\Lambda = 0$ there a single degenerated critical point at the origin. For $\Lambda > 0$ there are two critical points, unstable deS$_-$ and stable deS$_-$.

$x = 0$ and it is a nonhypebolic critical point since $Df(0,0) = 0$; the vector field $f(x) = -x^2$ is structurally unstable; $\Lambda = 0$ is a bifurcation value. For $\Lambda < 0$ there are no critical points. The phase portraits for this differential equation are shown in Fig. 2.

In this case we have $W^s(\sqrt{\frac{\Lambda}{3}}) = (-\sqrt{\frac{\Lambda}{3}}, \infty)$ and $W^u(-\sqrt{\frac{\Lambda}{3}}) = (-\infty, \sqrt{\frac{\Lambda}{3}})$ as, respectively, a stable and unstable manifold . And for $\Lambda = 0$ the one–dimensional center manifold is given by $W^c(0) = (-\infty, \infty)$. All of the pertinent information concerning the bifurcation that takes place in this system at $\Lambda = 0$ is captured in the bifurcation diagram shown in Fig. 3. The curve $\frac{\Lambda}{3} - x^2 = 0$ determines the position of the critical points of the system, a solid curve is used to indicate a family of stable critical points while a dashed curve is used to indicate a family of unstable critical points. This type of bifurcation is called a saddle-mode bifurcation.

The system under consideration constitutes only an example of dynamical system analysis of the system cosmological origin but there are many other systems with some parameter which shows hidden and unexpected properties as parameter varies. Let us remind some of them. In the problem of the motion star around the elliptic galaxy appears Henon, Heiles [14] Hamiltonian system. This system possesses the energy first integral E and if $E > E_{crit}$ then transition to the chaotic behavior appears. Another example of bifurcation and emergence of the cyclic behavior in the system of the limit cycle type is offered by the famous van der Pol equation $\ddot{x} + \mu(x^2 - 1)\dot{x} + x = 0$. For $\mu = 0$ the system is of harmonic oscillator type and for $\mu > 0$, van der Pol's equation has a unique limit cycle and is stable [15]. The limit cycle is representing a closed trajectory in the phase space which attracts all trajectories from neighborhood.

In this case $\epsilon = 0$ is a bifurcation value parameter and limit cycle behavior is an upper-level emergent property. For the interesting discussion

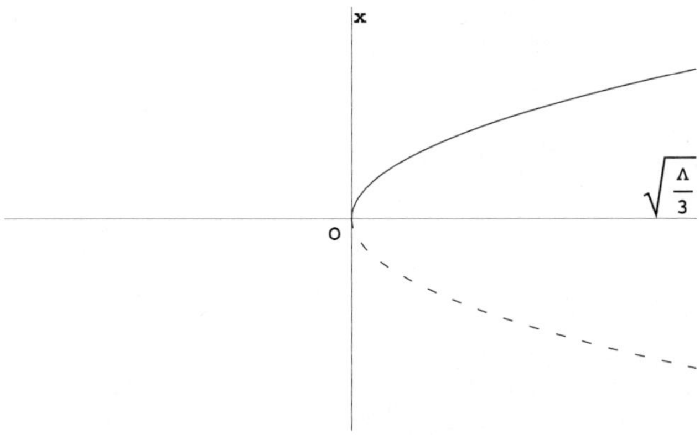

Figure 3: The bifurcation diagram of the flat LDM model (LCDM model without matter). It is the fork bifurcation of saddle-node type. The Einstein universe ($x = 0$) bifurcates to the expanding (upper stable branch) and contracting (lower unstable branch) de Sitter models.

on emergence, basal and upper-level models and reducibility see [16].

Also interesting experiences of emergence new type of dynamical behavior are provided by Hopf bifurcation phenomena [15, s. 341]. This bifurcation can occur in the system with parameter $\dot{x} = f(x, \mu)$ at a non-hyperbolic equilibrium point x_0 when the matrix $Df(x_0, \mu_0)$ has a simple pair of pure imaginary eigenvalue and no other eigenvalues with zero real point. In the generic case Hopf bifurcation occurs where a periodic orbit is created as the stability of equilibrium point x_μ changes. This type of behavior plays an important role in the description route to turbulence scenario. It would be worth mentioning the important role of Hopf bifurcation in Rulle-Takens scenario of route to deterministic chaos. The concept of turbulence war originally introduced by Landau in 1944 and later revised by Ruelle and Takens in 1941 [17]. According to Landau, turbulence is reached at the end of an indefinite superposition of oscillatory bifurcations, each bringing its unveiling phase into the dynamics of the system. In the Ruelle-Takens scenario infinite number of periodic behavior is not required when nonlinearities act. They argue that turbulence should be treated as a stochastic regime of deterministic chaos at which long term unpredictability occurs due to property of sensitive dependence on initial condition. This stage is reached only after a finite and small number of bifurcations. For some recent philosophical discussion on the significance of chaos see [18].

6. Conclusion

The main aim of this paper was to show the effectiveness of using the framework of dynamical system theories (especially the notion of structural stability and tools of bifurcation analysis) in the study on emergence and reducibility of two cosmological models Cold Dark Matter model and Lambda Cold Dark Matter model.

We have also demonstrated reducibility of solutions of LCDM upper state after taking the limit $\Lambda \to 0$. If we include the theorem about smooth dependence of solutions on initial condition and parameters, one can obtain CDM model as the boundary crossing. The analogical problem appears in the Wayne analysis of limit cycle behavior emergence in nonlinear system. In this case $\epsilon = 0$ is bifurcation value of parameter. In our interpretation space of state of the system can be parametrized by epsilon parameter which measures the strength of the nonlinear term. As a consequence bifurcation analysis reveals a new type of dynamical behavior for any value of epsilon parameter. This upper state can be reduced to the basal state because it is guaranteed by theorems mentioned above, which we applied in the context of LCDM model. We can find a strict analogy to the system under consideration of cosmological origin and Wayne analysis of emergence and singular limits. Following the common approach to reductionism in physics, so called deductive criterion of reducibility by Nagel (1961), the reduction is a derivational relation between upper-level and base-level theories. The structural instability of the models teaches us that one should not distinguish the derivational relation either on the level of basic equation and on the level of solutions. In the mathematical modeling of physical processes, we always try to convey the features of typical, garden-variety, dynamical systems. In mathematics the exceptional cases are more complicated and numerous, and they interrupt the physical discussion. Moreover, dynamicists share an opinion that such exceptional systems do not arise very often because they are not typical. The history of mathematical dynamics presents the search for generic properties. We would like to distinguish a class of phase portraits that are far simpler than the arbitrary ones. This program was achieved for dynamical systems on the plane by Peixoto due to the conception of structural stability introduced in 1934 by Andronov and Leontovich. The criteria for structural stability rely upon two supplementary notions: a perturbation of the phase portraits (or vector field) and the topological equivalence (homeomorphism of the state phase). A phase portrait has the property of structural stability if all sufficiently small perturbations of it have equivalent phase portraits. For example, one considers a center type of critical points, then the addition of perturbation pointing outward results in a point repellor which is not topologically equivalent to the center. This is a primary example of structurally unstable system. In the opposite case saddle type of critical point is structurally stable and the phase portrait does not change under small perturbation.

In this paper we define the class of FRW cosmological models filled by dark energy as two-dimensional dynamical systems of a Newtonian type. They are characterized through the single smooth effective potential function of the scale factor or redshift. Among these classes of models we distinguish typical (generic) and exceptional (nongeneric) cases with the help of structural stability notion and the Peixoto theorem. We find that the LCDM model is structurally stable as opposed to the CDM model. We demonstrate that this model represents a typical structurally stable perturbation of CDM one. Therefore, the transition from the CDM model of the Universe toward the LCDM one, which includes the effects of the cosmological constant, can be understood as an emergence of the model from the exceptional case to the generic one. This case represents a generic model in this sense that small changes of its right-hand sides do not change the global phase portraits. In terms of the potential, the second order differential equation, one can classify different models of cosmic acceleration. It is shown that models with the accelerating phase (following the deceleration) are natural and typical from the point of view of the dynamical systems theory combined with the notion of structural stability.

From the dynamical system theory we know that the solution of a dynamical system smoothly depends on the time, initial condition and the parameters. As a result we obtain that in any case as the model is represented by dynamical system there is a reduction from the upper-level model to the lower-level one by taking the mathematical limit with respect to the parameter. In the case of cosmology, the corresponding limit is obtained as the cosmological constant parameter goes to zero.

Acknowledgements

The work was supported by the grant NCN DEC-2013/09/B/ST2/03455.

References

[1] S. Perlmutter et al. (Supernova Cosmology Project), Astrophys. J. **517**, 565 (1999), *eprint*: astro-ph/9812133.

[2] A. G. Riess et al. (Supernova Search Team), Astrophys. J. **607**, 665 (2004), *eprint*: astro-ph/0402512.

[3] V. Lukash, Nuovo Cim. **122B**, 1411 (2007), *eprint*: 0712.3356.

[4] P. J. E. Peebles, Nuovo Cim. **122B**, 1035 (2007), *eprint*: 0712.2757.

[5] A. A. Andronov and L. S. Pontryagin, Dokl. Acad. Sci. URSS **14**, 247 (1937).

[6] R. Thom, *Stabilité structurelle et morphogénése* (Interédition, Paris, 1977).

[7] M. Szydlowski, M. Heller, and Z. Golda, Gen. Relat. Grav. **16**, 877 (1984).

[8] J. Losee, *A historical introduction to the philosophy of science* (Oxford University Press, Oxford, 1993).

[9] G. Holton, in *Scientific Explanation*, edited by A. F. Heath (Clarendon Press, Oxford, 1981).

[10] S. S. Kokarev (2008), *eprint*: 0810.5080.

[11] M. Peixoto, Topology **1**, 747 (1962).

[12] P. J. E. Peebles, *Principles of Physical Cosmology* (Princeton University Press, Princeton, 1993).

[13] E. Nagel, *The Structure of science* (Hackett Publishing Company, Incorporated, New York, 1961).

[14] M. Henon and C. Heiles, Astron. J. **69**, 73 (1964).

[15] L. Perko, *Differential Equations and Dynamical Systems (Texts in Applied Mathematics, Vol 7)* (Springer-Verlag New York, New York, 2001).

[16] A. Wayne (2008), extiteprint: http://philsci-archive.pitt.edu/archive/00003933.

[17] P. Cvitanovic, in *Universality in Chaos, 2nd ed*, edited by P. Cvitanovic (Taylor & Francis, 1989).

[18] C. Werndl (2008), *eprint*: http://philsci-archive.pitt.edu/archive/00003914.

[19] A. Kurek and M. Szydlowski, Astrophys. J. **675**, 1 (2008), *eprint*: astro-ph/0702484.

[20] R. C. Bishop and H. Atmanspacher, Found. Phys. **36** **12**, 1757 (2006).

Killing forms on homogeneous Sasaki-Einstein manifold $T^{1,1}$ *

Mihai Visinescu†
Department Theoretical Physics
National Institute for Physics and Nuclear Engineering
Magurele, Bucharest, Romania

ABSTRACT

We describe the construction of Killing forms on Sasaki-Einstein spaces. Some of Killing forms are related to the contact 1-form of the Sasaki manifolds. Two additional Killing forms are associated with the complex holomorphic volume form of the Calabi-Yau metric cones of the Sasaki-Einstein manifolds. We exemplify the general scheme in the case of the five dimensional homogeneous Sasaki-Einstein space $T^{1,1}$.

PACS: 04.62.+v, 04.65.+e

MSC2000: 83C40; 83C47

1. Introduction

In the last time Sasaki-Einstein geometry has become of significant interest in some modern developments in mathematics and theoretical physics [1, 2]. To understand the importance of Sasakian geometry we recall the fact that it sits naturally in between two Kähler geometries. On the one hand metric cones of the Sasakian manifolds are Kähler. Moreover, in the case of the Sasaki-Einstein spaces the metric cones are Ricci flat, i.e. Calabi-Yau manifolds. On the other hand any Sasakian manifold is contact and the one-dimensional foliation associated to the characteristic Reeb vector field is transversally Kähler [3].

The aim of this paper is the investigation of the hidden symmetries on toric Sasaki-Einstein manifolds. For this purpose we construct the (conformal) Killing forms using the interplay between complex coordinates of the Calabi-Yau cone and the special Killing forms on the toric Sasaki-Einstein space. We exemplify the general scheme in the case of the five dimensional $T^{1,1}$ space. The $T^{1,1}$ manifold is the first example of toric

* The work has been partly supported by the project *Cosmology and Strings PRJ-09* and partly by the project *JINR-Dubna 05-6-1119-2014/2016.*

† e-mail address: mvisin@theory.nipne.ro

Sasaki-Einstein/quiver duality [4]. The explicit homogeneous $T^{1,1}$ metric was constructed in [5].

The paper is organized as follows. In the next Section we describe the relation between the Sasaki-Einstein manifolds and their Calabi-Yau cones. In Section 3 we give some definitions regarding the (conformal) Killing forms and the particular class of special Killing forms. We present the complete list of special Killing forms on Sasaki-Einstein manifolds. In Section 4 we exemplify the construction of the special Killing forms on the five-dimensional $T^{1,1}$ space. The paper ends with conclusions in Section 5.

2. Sasaki-Einstein geometry

A $(2n-1)$-dimensional manifold M is a *contact manifold* if there exists a 1-form η (called a *contact 1-form*) on M such that

$$\eta \wedge (d\eta)^{n-1} \neq 0\,. \tag{1}$$

The *Reeb vector field* ξ dual to η ($\eta = \xi^*$) satisfies

$$\eta(\xi) = 1 \quad \text{and} \quad \xi \lrcorner d\eta = 0\,. \tag{2}$$

A contact Riemannian manifold (M, g_M) is Sasakian if its metric cone $C(M)$

$$C(M) \cong \mathbb{R}_+ \times M\,,\; g_{C(M)} = dr^2 + r^2\, g_M\,, \tag{3}$$

is Kähler [1] with Kähler form [6, 7]

$$\omega = \frac{1}{2} d(r^2\eta) = r dr \wedge \eta + \frac{1}{2} r^2 d\eta\,. \tag{4}$$

Here $r \in (0, \infty)$ may be considered as a coordinate on the positive real line $\mathbb{R}_+$. The Sasakian manifold (M, g_M) is naturally isometrically embedded into the metric cone via the inclusion

$$M = \{r = 1\} = \{1\} \times M \subset C(M)\,. \tag{5}$$

In the case of Sasaki-Einstein manifolds there is a well known fact that the following statements are equivalent [8]:

(i) (M, g_M) is Sasaki-Einstein with $\mathrm{Ric}\, g_M = 2(n-1) g_M$;

(ii) The Kähler cone $(C(M), g_{C(M)})$ is Ricci-flat ($\mathrm{Ric}\, g_{C(M)} = 0$), i.e. Calabi-Yau manifold;

(iii) The transverse Kähler structure to the Reeb foliation $\mathcal{F}_\xi$ is Kähler-Einstein with $\mathrm{Ric}\, g^T = 2n g^T$.

There are many mathematical and physical reasons to consider toric Sasaki-Einstein manifolds. For example, for $n = 3$ a toric Sasaki-Einstein manifold is dual to a *toric quiver gauge theory*.

Referring to metric cone, $C(M)$ is toric if the standard n-torus $\mathbb{T}^n = \mathbb{R}^n / 2\pi\mathbb{Z}^n$ acts effectively on it, preserving the Kähler form ω.

3. Killing forms

The *conformal Killing-Yano tensors* provide a natural generalization of Killing vector fields. Conformal Killing-Yano tensors are sometimes referred as *twistor forms* or *conformal Killing forms.*

A conformal Killing-Yano tensor of rank p on a n-dimensional Riemannian manifold (M, g_M), endowed with the Levi-Civita connection ∇, is a p-form Ψ which satisfies

$$\nabla_X \Psi - \frac{1}{p+1} X \lrcorner d\Psi + \frac{1}{n-p+1} X^* \wedge d^* \Psi = 0\,, \tag{6}$$

for any vector field X on M.

Here d and d^* are the de Rham differential and co-differential operators with respect to the Riemannian structure g_M, $\lrcorner$ is the operator dual to the wedge product $\wedge$ and X^* is the 1-form dual to the vector field X.

In component notation, the conformal Killing-Yano tensor equation is given by

$$\begin{aligned}\nabla_{(i_1}\psi_{i_2)i_3\cdots i_{p+1}} = \frac{1}{n-p+1}\Big(& g_{i_1 i_2}\nabla_j \psi^j{}_{i_3\cdots i_{p+1}} \\ & -(p-1) g_{[i_3(i_1}\nabla_j \psi^j{}_{i_2)i_4\cdots i_{p+1}]}\Big)\,.\end{aligned} \tag{7}$$

We used round and square brackets, respectively, to denote symmetrization or antisymmetrization over the indices within.

Coclosed conformal Killing forms are called *Killing forms.* If $p = 1$, the Killing forms are just dual to Killing vector fields.

A particular class of Killing forms is represented by the *special Killing forms*, which satisfy, for some constant c, the equation [9]

$$\nabla_X (d\Psi) = c X^* \wedge \Psi\,, \tag{8}$$

for any vector field X on M. It is worth mentioning the fact that the most known Killing forms are actually special.

Having in mind our interest to describe the Killing forms on Sasaki-Einstein spaces, we note the existence of a correspondence between special Killing forms defined on the Sasaki-Einstein manifold M and the parallel forms defined on its metric cone $C(M)$. More exactly, a p-dimensional differential form Ψ is a special form on M if and only if the corresponding form

$$\Psi_{cone} := r^p dr \wedge \Psi + \frac{r^{p+1}}{p+1} d\Psi\,, \tag{9}$$

is parallel on $C(M)$.

On a $(2n-1)$-dimensional Sasaki manifold with the contact 1-form η there are the following special Killing forms:

$$\Psi_k = \eta \wedge (d\eta)^k \quad , \quad k = 0, 1, \cdots, n-1\,. \tag{10}$$

Indeed they are special Killing forms (8) as Ψ_k satisfies for any vector field X and any k the additional equation [9]

$$\nabla_X (d\Psi_k) = -2(k+1)X^* \wedge \Psi_k\,. \tag{11}$$

Besides these Killing forms, there are $n-1$ closed conformal Killing forms (also called $*$-Killing forms)

$$\Phi_k = (d\eta)^k \quad , \quad k = 1, \cdots, n-1\,. \tag{12}$$

Let us note that in the case of the Calabi-Yau cone the holonomy is $SU(n)$ and there are *two additional* parallel forms of degree n. In order to write explicitly the additional Killing forms which correspond to these parallel forms, we shall express the volume form of the metric cone in terms of the Kähler form (4)

$$d\mathcal{V} = \frac{1}{n!}\omega^n\,. \tag{13}$$

Here ω^n is the wedge product of ω with itself n times. The volume of a Kähler manifold can be also written as [9, 10]

$$d\mathcal{V} = \frac{i^n}{2^n}(-1)^{n(n-1)/2}\Omega \wedge \overline{\Omega}\,, \tag{14}$$

where Ω is the complex volume holomorphic $(n,0)$ form of $C(M)$. The additional (real) parallel forms are given by the real and respectively imaginary part of the complex volume form.

4. Special Killing forms on $C(T^{1,1})$ space

There are known various compactifications of chiral $\mathcal{N}=2$ ten-dimensional supergravity to five dimensions. The simplest compactification, apart from S^5, involve the coset spaces $T^{p,q}$ [11]. Of these, only $T^{1,1}$ preserves some supersymmetry, namely this compactification should be dual to an $\mathcal{N}=1$ superconformal field theory in four dimensions. $T^{1,1}$ is a homogeneous space $T^{1,1} = (SU(2) \times SU(2)/U(1)$ with the $U(1)$ being a diagonal subgroup of the maximal torus of $SU(2) \times SU(2)$.

The Calabi-Yau cone $C(T^{1,1})$ over the homogeneous Sasaki-Einstein manifold $T^{1,1}$ can be described by the following equation in four complex variables $z_i \in \mathbb{C}$

$$z_1^2 + z_2^2 + z_3^2 + z_4^2 = 0\,, \tag{15}$$

with a double point singularity at $z_i = 0$. The manifold $T^{1,1}$ can be identified as the intersection of the conifold (15) and the sphere

$$\sum_{i=1}^{4} |z_i|^2 = 1\,. \tag{16}$$

The holomorphic three form of the Calabi-Yau cone $C(T^{1,1})$ is [4]:

$$\Omega = \frac{dz_2 \wedge dz_3 \wedge dz_4}{z_1}\,, \tag{17}$$

and the metric on the conifold may be written as

$$ds^2_{C(T^{1,1})} = dr^2 + r^2 ds^2_{T^{1,1}}\,. \tag{18}$$

$T^{1,1}$ is a S^1 bundle over $S^2 \times S^2$ and the metric may be written as [5, 6]

$$\begin{aligned} ds^2(T^{1,1}) =& \frac{1}{6}(d\theta_1^2 + \sin^2\theta_1 d\phi_1^2 d\theta_2^2 + \sin^2\theta_2 d\phi_2^2) \\ &+ \frac{1}{9}(d\psi + \cos\theta_1 d\phi_1 + \cos\theta_2 d\phi_2)^2\,. \end{aligned} \tag{19}$$

The coordinates (θ_1, ϕ_1) and (θ_2, ϕ_2) parametrize the sphere S^2 in a standard way and the angle $\psi \in [0, 4\pi)$ parametrizes the $U(1)$ fiber.

We have three commuting $U(1)$ actions preserving the Kähler form of $C(T^{1,1})$ generated by $\partial/\partial\phi_i$, $i = 1, 2$ and $\partial/\partial\psi$. In particular $3\partial/\partial\psi$ is the Reeb vector so that $T^{1,1}$ is a regular Sasaki-Einstein manifold and its base Kähler-Einstein manifold is just $C\mathbb{P}^1 \times C\mathbb{P}^1$ [6].

The symplectic form on the cone is

$$\begin{aligned} \omega =& -\frac{r^2}{6}(\sin\theta_1 d\theta_1 \wedge d\phi_1 + \sin\theta_2 d\theta_2 \wedge d\phi_2) \\ &+ \frac{r}{3} dr \wedge (d\psi + \cos\theta_1 d\phi_1 + \cos\theta_2 d\phi_2)\,. \end{aligned} \tag{20}$$

Taking into account (4) we could extract the contact 1-form η of the symplectic manifold $T^{1,1}$.

$$\eta = \frac{1}{3}(d\psi + \cos\theta_1 d\phi_1 + \cos\theta_2 d\phi_2)\,, \tag{21}$$

and

$$d\eta = -\frac{1}{3}(\sin\theta_1 d\theta_1 \wedge d\phi_1 + \sin\theta_2 d\theta_2 \wedge d\phi_2)\,. \tag{22}$$

Using the explicit expression of the contact 1-form η we can proceed directly to the construction of the special Killing forms Ψ_k (10) and conformal Killing forms Φ_k (12).

Two additional special Killing 2-forms are associated with the complex holomorphic volume form Ω (17) of the Calabi-Yau metric cone. For this purpose we express the complex coordinates z_i in terms of the coordinates $(r, \theta_i, \phi_i, \psi)$ as follows [12]:

$$\begin{aligned}
z_1 =& \frac{r^{3/2} e^{i\psi/2}}{\sqrt{2}} \\
&\times \left[\cos\left(\frac{\theta_1+\theta_2}{2}\right)\cos\left(\frac{\phi_1+\phi_2}{2}\right) + i\cos\left(\frac{\theta_1-\theta_2}{2}\right)\sin\left(\frac{\phi_1+\phi_2}{2}\right)\right] , \\
z_2 =& -\frac{r^{3/2} e^{i\psi/2}}{\sqrt{2}} \\
&\times \left[\cos\left(\frac{\theta_1+\theta_2}{2}\right)\sin\left(\frac{\phi_1+\phi_2}{2}\right) - i\cos\left(\frac{\theta_1-\theta_2}{2}\right)\cos\left(\frac{\phi_1+\phi_2}{2}\right)\right] , \\
z_3 =& -\frac{r^{3/2} e^{i\psi/2}}{\sqrt{2}} \\
&\times \left[\sin\left(\frac{\theta_1+\theta_2}{2}\right)\cos\left(\frac{\phi_1-\phi_2}{2}\right) - i\sin\left(\frac{\theta_1-\theta_2}{2}\right)\sin\left(\frac{\phi_1-\phi_2}{2}\right)\right] , \\
z_4 =& -\frac{r^{3/2} e^{i\psi/2}}{\sqrt{2}} \\
&\times \left[\sin\left(\frac{\theta_1+\theta_2}{2}\right)\sin\left(\frac{\phi_1-\phi_2}{2}\right) + i\sin\left(\frac{\theta_1-\theta_2}{2}\right)\cos\left(\frac{\phi_1-\phi_2}{2}\right)\right] .
\end{aligned} \tag{23}$$

For later use it is convenient to introduce the coordinates $\theta_\pm$ and $\phi_\pm$ instead of θ_i and ϕ_i:

$$\theta_\pm = \frac{\theta_1 \pm \theta_2}{2} \quad , \quad \phi_\pm = \frac{\phi_1 \pm \phi_2}{2} . \tag{24}$$

In terms of these coordinates, the holomorphic form (17) becomes:

$$\begin{aligned}
\Omega =& \frac{r^3 e^{i\psi}}{2} \frac{[cos\theta_+ \cos\phi_+ - i\cos\theta_- \sin\phi_+]}{[\cos^2\theta_+ \cos^2\phi_+ + \cos^2\theta_- \sin^2\phi_+]} \\
&\times \left[\frac{3}{2}\frac{dr}{r}(-\cos\theta_+ \sin\phi_+ + i\cos\theta_- \cos\phi_+)\right. \\
&+\frac{i}{2}d\psi(-\cos\theta_+ \sin\phi_+ + i\cos\theta_- \cos\phi_+) \\
&+(\sin\theta_+ \sin\phi_+ d\theta_+ - \cos\theta_+ \cos\phi_+ d\phi_+) \\
&\left.+i(-\sin\theta_- \cos\phi_+ d\theta_- - \cos\theta_- \sin\phi_+ d\phi_+)\right]
\end{aligned}$$

$$\begin{aligned}
&\wedge \left[\frac{3}{2}\frac{dr}{r}(-\sin\theta_+\cos\phi_- + i\sin\theta_-\sin\phi_-)\right.\\
&+\frac{i}{2}d\psi(-\sin\theta_+\cos\phi_- + i\sin\theta_-\sin\phi_-)\\
&+(-\cos\theta_+\cos\phi_- d\theta_+ + \sin\theta_+\sin\phi_- d\phi_-)\\
&\left.+i(\cos\theta_-\sin\phi_- d\theta_- + \sin\theta_-\cos\phi_- d\phi_-)\right]\\
&\wedge \left[\frac{3}{2}\frac{dr}{r}(-\sin\theta_+\sin\phi_- - i\sin\theta_-\cos\phi_-)\right.\\
&+\frac{i}{2}d\psi(-\sin\theta_+\sin\phi_- - i\sin\theta_-\cos\phi_-)\\
&+(-\cos\theta_+\sin\phi_- d\theta_+ - \sin\theta_+\cos\phi_- d\phi_-)\\
&\left.+i(-\cos\theta_-\cos\phi_- d\theta_- + \sin\theta_-\sin\phi_- d\phi_-)\right].
\end{aligned}\tag{25}$$

In accordance with (9) to this 3-form Ω on the Calabi-Yau metric cone it corresponds a 2-form Ψ on the Sasaki-Einstein space $T^{1,1}$:

$$\Omega = r^2 dr\wedge\Psi + \frac{r^3}{3}d\Psi\,.\tag{26}$$

After a straightforward calculation we obtain:

$$\begin{aligned}
\Psi =&\frac{3}{4}e^{i\psi}\frac{[cos\theta_+\cos\phi_+ - i\cos\theta_-\sin\phi_+]}{[\cos^2\theta_+\cos^2\phi_+ + \cos^2\theta_-\sin^2\phi_+]}\\
&\times[-(\cos\theta_+\cos\phi_+ + i\cos\theta_-\sin\phi_+)d\theta_+\wedge d\theta_-\\
&+\cos\theta_+\sin\theta_-(-\cos\theta_-\sin\phi_+ + i\cos\theta_+\cos\phi_+)d\theta_+\wedge d\phi_+\\
&+\sin\theta_+\cos\theta_-(-\cos\theta_-\sin\phi_+ + i\cos\theta_+\cos\phi_+)d\theta_+\wedge d\phi_-\\
&+\sin\theta_+\cos\theta_-(\cos\theta_-\sin\phi_+ - i\cos\theta_+\cos\phi_+)d\theta_-\wedge d\phi_+\\
&+\cos\theta_+\sin\theta_-(\cos\theta_-\sin\phi_+ - i\cos\theta_+\cos\phi_+)d\theta_-\wedge d\phi_-\\
&+(\sin^2\theta_+ - \sin^2\theta_-)(\cos\theta_+\cos\phi_+ + i\cos\theta_-\sin\phi_+)d\phi_+\wedge d\phi_-]\,.
\end{aligned}\tag{27}$$

From here, calculating the real and imaginary part of Ψ we get two new special Killing 2-forms.

5. Conclusions

The importance of Killing forms comes from their role in physics being related to hidden symmetries, superintegrability of field equations, quantum symmetry operators, supersymmetries, etc.

In general it is a hard task to solve the generalized Killing equation (6). In addition to the brute force method of trying to solve (6) in some cases there are other methods which are considerable simpler. In the case of Sasaki-Einstein spaces the complete set of special Killing forms is related to the geometrical structures. Special Killing forms are constructed from the contact 1-form of the Sasaki-Einstein spaces and other additional Killing forms are associated with the complex volume form of the Calabi-Yau metric cone.

In this paper we give a method to construct special Killing forms on a toric Sasaki geometry and exemplify the general scheme in the case of the five dimensional homogeneous Sasaki-Einstein manifold $T^{1,1}$. Many examples of Sasaki-Einstein manifolds may be obtain via toric geometry [7, 13, 14, 15], and such examples are a good testing ground for the AdS/CFT correspondence [16].

References

[1] C. P. Boyer and K. Galicki, *Sasakian geometry*, Oxford Mathematical Monographs, Oxford University Press, Oxford, 2008.

[2] J. Sparks, *Surv. Diff. Geom.* **16** (2011) 265.

[3] C. P. Boyer, K. Galicki and S. R. Simanca, *Comm. Math. Phys.* **279** (2008) 705.

[4] I. R. Klebanov and E. Witten, *Nucl. Phys.* **B536** (1998) 199.

[5] P. Candelas and X. de la Ossa, *Nucl. Phys.* **B342** (1990) 246.

[6] D. Martelli and J. Sparks, *Commun. Math. Phys.* **262** (2006) 51.

[7] D. Martelli, J. Sparks and S.-T. Yau, *Commun. Math. Phys.* **268** (2006) 39.

[8] A. Futaki, H. Ono and G. Wang, *J. Diff. Geom.* **83** (2009) 585.

[9] U. Semmelmann, *Math. Z.* **245** (2003) 503.

[10] M. Visinescu and G. E. Vîlcu, *SIGMA* **8** (2012) 058.

[11] L. J. Romans, *Phys. Lett.* **153B** (1985) 392.

[12] S. S. Gubser and I. R. Klebanov, *Phys. Rev.* **D58** (1998) 125025.

[13] T. Oota and Y. Yasui, *Nucl. Phys.* **B742** (2006) 275.

[14] M. Visinescu, *Mod. Phys. Lett. A* **27** (2012) 1250217.

[15] V. Slesar, M. Visinescu and G. E. Vîlcu, *Phys. Scr.* **89** (2014) 125205.

[16] J. M. Maldacena, *Adv. Theor. Math. Phys.* **2** (1998)231.

Lie symmetries of the Wheeler-DeWitt equation with application in Hybrid Gravity

Andrzej Borowiec*
Institute for Theoretical Physics, pl. M. Borna 9, 50-204,Wroclaw, Poland
Salvatore Capozziello†
Dipartimento di Fisica, Università di Napoli "Federico II", Napoli, Italy
Andronikos Paliathanasis‡
Dipartimento di Fisica, Università di Napoli "Federico II", Napoli, Italy
Aneta Wojnar¶
Institute for Theoretical Physics, pl. M. Borna 9, 50-204, Wroclaw, Poland.

Abstract

Lie point symmetries for Hybrid Gravity models are considered in order to select potentials and to find invariant solutions of the Wheeler-DeWitt equation. Analytical solution of the modified field equations determines the Hubble function H^2 as a fourth order polynomial with five non-vanishing coefficients. It can be reinterpreted in terms of well-known, from the extended Standard Cosmological Model (ΛCDM), perfect fluid components.

1. Introduction

Although General Relativity (GR) is a beautiful theory verified by numerous astronomical observations, there are still phenomena such as late-time cosmic acceleration [1, 2, 3], dark matter and dark energy [4] that the standard theory is not able to explain. It seems that some modifications of GR are necessary to answer that questions. One of them is $f(R)$ gravity [5, 6] which instead of the Ricci scalar R in the Hilbert-Einstein Lagrangian takes an unknown function of R. Such a consideration can be treated in two formalisms: metric and Palatini one (see e.g. [7] and references there in) where in the former the metric and connection are considered as two independent variables and curvature $\mathcal{R}$ is constructed of the

* e-mail address: andrzej.borowiec@ift.uni.wroc.pl
† e-mail address: capozzie@na.infn.it
‡ e-mail address: paliathanasis@na.infn.it
§ AW would like to thank the Organizers for warm hospitality during the Conference. AW acknowledges financial support from the NCN project DEC-2013/09/B/ST2/03455
¶ e-mail address: aneta.wojnar@ift.uni.wroc.pl

connection. Both formalisms can be combined in the approach denoted as hybrid metric-Palatini gravity or $f(X)$ gravity [8, 9], where the action is taken as the standard Hilbert-Einstein case (linear in Ricci scalar R) with a nonlinear term in the Palatini curvature scalar $\mathcal{R}$. The equations of motions and Wheeler-DeWitt (WDW) equation of that theory can be examined by using point symmetries which help to solve the given differential equations. It is also possible to find potentials appearing in the model which are distinguished by Noether symmetries. Here, we will present the Hybrid Gravity model in spatially flat Friedmann-Robertson-Walker spacetime with solutions of the classical field equations and invariant solutions of WDW equation.

2. Symmetries of differential equations

A Lie symmetry of a differential equation is a one parameter point transformation which leaves the differential equation invariant. Lie symmetries is the main tool to study nonlinear differential equations. Indeed, Lie symmetries provide invariant functions which can be used to construct analytic solutions for a differential equation. For differential equations which arise from a variational principle there may exist a special class of Lie symmetries which are called Noether symmetries. Noether symmetries are the generators of the one parameter point transformation which transform the Lagrangian such as the Euler-Lagrange equation to be invariant. The importance of Noether symmetries is that they provide use of first integrals/conservation laws [10].

Consider the second order partial differential equation (PDE) $\mathcal{H} \equiv \mathcal{H}(x^i, u^A, u^B_{,i}, u^C_{,ij}) = 0$ where x^i are the independent variables and u^A are the dependent ones. We shall say that the PDE $\mathcal{H}$ is invariant under the action of the infinitesimal point transformation

$$\bar{x}^i = x^i + \varepsilon\xi^i(x^k, u^B)\ ,\ \bar{u}^A = u^A + \varepsilon\eta^A(x^k, u^B) \tag{1}$$

with infinitesimal generator

$$\mathbf{X} = \xi^i(x^k, u^B)\partial_{x^i} + \eta^A(x^k, u^B)\partial_{u^A} \tag{2}$$

if and only if there exists a function λ such that the following condition holds [11]

$$\mathbf{X}^{[2]}(\mathcal{H}) = \lambda\mathcal{H}\ ,\ mod\mathcal{H} = 0, \tag{3}$$

where $\mathbf{X}^{[2]} = \mathbf{X} + \eta_i^A\partial_{u_i^A} + \eta_{ij}^A\partial_{u_{ij}^A}$ is the second prolongation vector of $\mathbf{X}$ and[1]

$$\eta_i^A = D_i\eta^A - u_j^A D_i\xi^j,\ \ \eta_{ij}^A = D_i\eta_j^A - u_{ik}^A D_j\xi^k. \tag{4}$$

For PDEs arising from a variational principle, Noether's theorem states [12] that the action of the generator (2) of the infinitesimal transformation (1) acting on the Lagrangian $L = L(x^k, u, u_k)$ leaves a field equation

[1]Where $D_i = \partial_i + u_i\partial_u + u_{ij}\partial_{u_j} + ...$ is the total derivative.

$\mathcal{H}(x^i, u, u_{,i}, u_{,ij}) = 0$ invariant if there exists a vector field $A^i = A^i(x^i, u)$ such that the following condition is satisfied ($\mathbf{X}^{[1]} = \mathbf{X} + \eta_i \partial_{u_i}$):

$$\mathbf{X}^{[1]} L + L D_i \xi^i = D_i A^i \ . \tag{5}$$

The corresponding Noether flow I^i is defined by the expression

$$I^i = \xi^k \left(u_k \frac{\partial L}{\partial u_i} - L \right) - \eta \frac{\partial L}{\partial u_i} + A^i \tag{6}$$

being conserved if the relation $D_i I^i = 0$ is satisfied.

The method of the Noether symmetries of the field has been applied by many authors as a selection rule of the unknown functions in various cosmological models, e.g. in scalar field cosmology, in f(R) gravity and others (for instance see [13, 14, 15, 16, 17, 18, 19]). However in [20, 21] it has been shown that this selection rule has geometrical properties since the Noether symmetries are related with the collineations of the minisuperspace. Recently, in [22] it has been proposed that the cosmological model will be determined by the existence of Lie symmetries of the Wheeler-DeWitt equations (WDW) in quantum cosmology. This selection rule is more general and it is related with the Noether symmetry criterion.

In the following we will apply this method in order to determine exact solutions of the WDW equation by using the Lie symmetries in Hybrid Gravity. Furthermore, we will show that when the WDW equation is separable the classical field equations are integrable and vice versa.

3. Hybrid Gravity

The action of the hybrid metric-Palatini gravity has the following form [8, 9]:

$$S = \frac{1}{2\kappa^2} \int \mathrm{d}^4 x \sqrt{-g} [R + f(\mathcal{R})] + S_m \tag{7}$$

where R is the metric curvature scalar and $f(\mathcal{R})$ is a function of the Palatini curvature scalar which is constructed by an independent torsionless connection $\hat{\Gamma}$. One uses the metric g in order to construct the scalar from the Ricci tensor of $\hat{\Gamma}$, i.e. $\mathcal{R} = g^{\mu\nu} \mathcal{R}_{\mu\nu}(\hat{\Gamma})$. The variation of the above action with respect to the connection gives modified gravitational field equations

$$G_{\mu\nu} + f'(\mathcal{R}) \mathcal{R}_{\mu\nu} - \frac{1}{2} f(\mathcal{R}) g_{\mu\nu} = \kappa^2 T_{\mu\nu}, \tag{8}$$

where $G_{\mu\nu}$ is the Einstein tensor for the metric $g_{\mu\nu}$ (with Lorentzian signature). It can be shown that $R_{\mu\nu}(g)$ and $\mathcal{R}_{\mu\nu}(h)$ are conformally related to each other by the relation $h_{\mu\nu} = f'(\mathcal{R}) g_{\mu\nu}$, where the conformal factor is given by $f'(\mathcal{R}) = df(\mathcal{R})/d\mathcal{R}$. The trace of Eq. (8) is the hybrid structural equation, where one can algebraically express the Palatini curvature $\mathcal{R}$ in terms of a quantity X assuming that the following equation has analytic solutions:

$$f'(\mathcal{R})\mathcal{R} - 2f(\mathcal{R}) = \kappa^2 T + R \equiv X. \tag{9}$$

The variable X measures the deviation from the General Relativity trace equation $R = -\kappa^2 T$, that is GR is fully recovered for $X = 0$ [9].

As for the pure metric and Palatini case [5, 7], the above action can be transformed into a scalar-tensor theory if $f''(\mathcal{R}) \neq 0$

$$S = \frac{1}{2\kappa^2}\int \mathrm{d}^4x\sqrt{-g}[R + \phi\mathcal{R} - V(\phi)]. \tag{10}$$

where $\phi \equiv f'(\mathcal{R})$ and $V(\phi) = \mathcal{R}f'(\mathcal{R}) - f(\mathcal{R})$. Using the relation between R and $\mathcal{R}$, given by $\mathcal{R} = R + \frac{3}{2\phi^2}\partial_\mu\phi\partial^\mu\phi - \frac{3}{\phi}\Box\phi$ (see [9] for details) one finally obtains the standard scalar-tensor form

$$S = \frac{1}{2\kappa^2}\int \mathrm{d}^4x\sqrt{-g}\left[(1+\phi)R + \frac{3}{2\phi}\partial^\mu\phi\partial_\mu\phi - V(\phi)\right]. \tag{11}$$

Let us consider the FRW spatially flat metric

$$ds^2 = -N^2(t)\,dt^2 + a^2(t)\left(dx^2 + dy^2 + dz^2\right). \tag{12}$$

Therefore, from action (11) one deduces the pointlike Lagrangian

$$\mathcal{L} = \frac{1}{N}\left(6a\dot{a}^2(1+\phi) + 6a^2\dot{a}\dot{\phi} + \frac{3}{2\phi}a^3\dot{\phi}^2\right) + Na^3V(\phi). \tag{13}$$

The energy condition resulting from the above Lagrangian $\left(\frac{\partial L}{\partial N} = 0\right)$ is

$$\frac{1}{N^2}\left(6a\dot{a}^2(1+\phi) + 6a^2\dot{a}\dot{\phi} + \frac{3}{2\phi}a^3\dot{\phi}^2\right) - a^3V(\phi) = 0. \tag{14}$$

The details concerning the Friedmann and field equations of (13) can be find in [23].

From Lagrangian (13) we define the momentum p_a, p_ϕ as follows

$$p_a = \frac{6a}{N}\left(2(1+\phi)\dot{a} + a\dot{\phi}\right), \quad p_\phi = \frac{3a^2}{\phi N}\left(2\phi\dot{a} + a\dot{\phi}\right) \tag{15}$$

hence the energy condition (14) becomes

$$\frac{1}{12a}p_a^2 - \frac{\phi}{3a^2}p_a p_\phi + \frac{\phi(1+\phi)}{3a^3}p_\phi^2 - a^3V(\phi) = 0. \tag{16}$$

and the rest of the field equations are the Hamilton's equations of (16).

4. Symmetries of the WDW equation

Roughly speaking, WDW equation is a quantized version of a Hamiltonian of a considered system. In our case, since the minisuperspace is two-dimensional described by the minisuperspace metric G_{ij}, the WDW it has the following form

$$\Delta\Psi - a^3 V(\phi)\Psi = 0, \tag{17}$$

where $\Delta = \frac{1}{\sqrt{|G|}}\frac{\partial}{\partial x^i}\left(\sqrt{|G|}G^{ij}\frac{\partial}{\partial x^i}\right)$ is the Laplace operator, Ψ is a wave function of the Universe and $x^i = \{a, \phi\}$. Using Lie symmetries for the Lagrangian (13) one can find the form of the potential which in that case is a constant one [23].

Since the dynamical system (13) is conformally invariant, we can apply the conformal transformation $\mathrm{d}\tau = N(a)\mathrm{d}t$ to the system and find new solutions [24, 25, 26]. Therefore, one considers the conformal Lagrangian

$$\mathcal{L}\left(x^i, (x^i)'\right) = \frac{a^3 V(\phi)}{N(a)} + N(a)\left[6a(1+\phi)a'^2 + 6a^2 a'\phi' + \frac{3}{2\phi}a^3\phi'^2\right], \tag{18}$$

where the prime denotes $d/d\tau$. In order to simplify the problem we chose $N(a) = a^{-1}e^{N_0 a}$ for which we found two extra Noether symmetries with potentials (V_0 and V_1 are constant):

$$X_1 = -\frac{1}{2}\partial_a + \frac{\phi + V_1\sqrt{\phi}}{a}\partial_\phi, \quad V(\phi) = V_0\left(\sqrt{\phi} + V_1\right)^4; \tag{19}$$

$$X_2 = 2\tau\partial_\tau + a\left(\sqrt{\phi}V_1 + 1\right)\partial_a - 2V_1\sqrt{\phi}(\phi+1)\partial_\phi, \tag{20}$$

$$V(\phi) = V_0(1+\phi)^2 \exp\left(\frac{6}{V_1}\arctan\sqrt{\phi}\right). \tag{21}$$

The corresponding conservation laws are

$$I_1 = 6\left(V_1\sqrt{\phi} - 1\right)\dot{a} + 3\frac{a}{\sqrt{\phi}}V_1\dot{\phi}, \tag{22}$$

$$I_2 = 12a(1+\phi)\dot{a} + 6a^2\left(1 - \frac{V_1}{\sqrt{\phi}}\right)\dot{\phi}. \tag{23}$$

4.1. Exact solutions of WDW equation

In this paper we are focused on the case of power law potential (19). It has been shown [23] that after a coordinate transformation the Lagrangian (18) turns out to be

$$\mathcal{L}\left(x, y, x', y'\right) = \frac{1}{2}x'^2 + \frac{1}{2}y'^2 + \bar{V}_0 y^4, \tag{24}$$

where $\bar{V}_0$ is a constant and (x, y) are appropriate functions of (a, ϕ). The Hamiltonian of the field equations is given by

$$\tilde{H} = \frac{1}{2}p_x'^2 + \frac{1}{2}p_y'^2 - \bar{V}_0 y^4, \tag{25}$$

where p_x, p_y are the momenta. The WDW equation constructed from (25) reads

$$\Psi_{,xx} + \Psi_{,yy} - 2\bar{V}_0 y^4 \Psi = 0 \tag{26}$$

and it admits Lie point symmetries for the vector fields [26]

$$X_\Psi = c_1 \partial_x + (c_2 \Psi + b(x, y))\, \partial_\Psi, \quad X_b = b(x, y)\, \partial_\Psi, \tag{27}$$

where $b(x, y)$ is a function that satisfies (26). In order to reduce (26) and to find its solutions one may use the zeroth order invariants. The invariant solutions of the Wheeler-DeWitt equation is found to be

$$\Psi(x, y) = \sum_\mu \left[y_1 e^{\mu x + w(y)} + y_2 e^{\mu x - w(y)} \right], \tag{28}$$

where $w(y) = \frac{\sqrt{2}}{2} \int \sqrt{(2\bar{V}_0 y^4 - \mu^2)} \mathrm{d}y$ and $\mu \in \mathbb{C}$.

4.2. Classical solutions

Let us come back to the Hamiltonian (25) which has a constraint, i.e. $\tilde{H} = 0$. The Hamilton-Jacobi equation provides the action

$$S = c_1 x + \int \sqrt{2\bar{V}_0 y^4 - c_1^2} + S_0, \tag{29}$$

which allows us to reduce the field equations of (25) and find their exact solutions of the form

$$x(\tau) = x_1 \tau + x_2, \quad \int \frac{dy}{\sqrt{2\bar{V}_0 y^4 - c_1^2}} = \varepsilon(\tau - \tau_0), \tag{30}$$

where $\varepsilon = \pm 1$. Considering the case $V_1 = 0$, one has the solution $a(\tau) = a_3 \tau$ where $\tau = \sqrt{t}$ which is the radiation solution. Assuming $V_1 \neq 0$ and $c_1 = 0$, the scale factor reads instead

$$a(\tau) = a_1 - a_2 \frac{1}{\tau - \tau_0} + a_3(\tau - \tau_0), \tag{31}$$

where a_3, a_1, a_2, are constants. One obtains the Hubble function[2]:

$$H(a) = a'^2 = a_3 a^{-2} + 4a_3^2 a_2 \left(a^3 - a_1 a^2 + \varepsilon a^2 \sqrt{(a - a_1)^2 + 4a_3 a_2} \right)^{-2}, \tag{32}$$

[2]Recall that $H = \frac{1}{a}\frac{da}{dt} = \frac{1}{a^2}\frac{da}{d\tau}$

where $(a - a_1)^2 + 4a_3a_2 \geq 0$ and $a(\tau) \in \mathbb{R}$ must be satisfied for a real solution. Putting $a_3 = 0$ one finds that $a_2^{-1} = H_0/(|a_1|+1)^2$ so the Hubble function can be written as (at the present time $a_0 = 1$)

$$\frac{H^2(a)}{H_0^2} = \Omega_r a^{-4} + \Omega_m a^{-3} + \Omega_k a^{-2} + \Omega_f a^{-1} + \Omega_\Lambda, \tag{33}$$

where

$$\Omega_f = \frac{|4a_1|}{(|a_1|+1)^4}, \qquad \Omega_\Lambda = \frac{1}{(|a_1|+1)^4}, \qquad \Omega_r = \frac{|a_1|^4}{(|a_1|+1)^4},$$
$$\Omega_m = \frac{|4a_1|^3}{(|a_1|+1)^4}, \qquad \Omega_k = \frac{|6a_1|^2}{(|a_1|+1)^4}. \tag{34}$$

Each power of $\sqrt{\phi}$ in the potential above introduces a power of the scale factor in the Hubble function (33) of the model. It in turn allows perfect fluid interpretation as: radiation ($w = 1/3$), dust ($w = 0$), curvature-like fluid ($w = -1/3$), a (non-phantom) dark energy fluid ($w = -2/3$) and a cosmological constant ($w = -1$), where $w = p/\rho$ denotes the corresponding barotropic equation of state parameter. It should be noticed that we assumed spatially flat universe but the curvature-like term follows from the Hybrid Gravity.

5. Conclusions

The aim of the paper was to discuss applications of point symmetries in Hybrid Gravity model and the most important results from [23]. One needed to apply conformal transformation to the Lagrangian in order to obtain new classes of solutions which one of them was considered in the paper. For the power law potential $V(\phi) = V_0\left(\sqrt{\phi} + V_1\right)$ the Hubble function is a fourth-order polynomial with non-vanishing coefficients, i.e. each power of $\sqrt{\phi}$ produces a corresponding fluid in the model. The special form of the potential ($V_1 = 0$) gives radiation solutions. In that case it is possible to write WDW equation in a form allowing to find explicitly (using the zeroth order invariants) an invariant solution. This is Wave Function of the Universe [27] whose peaks appear for conserved momenta and observed universe (i.e. classical cosmological solution) [28].

References

[1] S. Perlmutter *et al.* [Supernova Cosmology Project Collaboration], Astrophys. J. **517**, 565 (1999).

[2] A. G. Riess *et al.* [Supernova Search Team Collaboration], Astron. J. **116**, 1009 (1998).

[3] A. G. Riess *et al.* [Supernova Search Team Collaboration], Astrophys. J. **607**, 665 (2004).

[4] S. Capozziello and M. De Laurentis, Annalen Phys. **524**, 545 (2012).

[5] S. Capozziello and M. De Laurentis, Phys. Rept. **509**, 167 (2011).

[6] S. Nojiri and S.D. Odintsov, Phys. Rept. 505, 59 (2011).

[7] G. J. Olmo, Int. J. Mod. Phys. D **20**, 413 (2011).

[8] T. Harko, T. S. Koivisto, F. S. N. Lobo and G. J. Olmo, Phys. Rev. D **85**, 084016 (2012).

[9] S. Capozziello, T. Harko, T. S. Koivisto, F. S. N. Lobo and G. J. Olmo, JCAP **1304**, 011 (2013).

[10] A. Paliathanasis, Symmetries of differential equations and applications in relativistic physics, (2014), PhD Thesis, University of Athens, Athens, Greece

[11] H. Stephani, Differential Equations: Their Solutions using Symmetry, Cambridge University Press, Cambridge (1989).

[12] G. Bluman and S. Kumei, Symmetries and differential equations, New York, Springer, 1989.

[13] N. Dimakis, T. Christodoulakis and P.A. Terzis, J.Geom.Phys. 77 (2014) 97-112

[14] B. Vakili and F. Khazaie, Class.Quant.Grav. 29 (2012) 035015

[15] S. Capozziello, N. Frusciante and D. Vernieri, Gen.Rel.Grav. 44 (2012) 1881-1891

[16] S. Capozziello and M. De Laurentis, Int.J.Geom.Meth.Mod.Phys. 11 (2014) 1460004

[17] S. Basilakos, S. Capozziello, M. De Laurentis, A. Paliathanasis and M. Tsamparlis, Phys.Rev. D88 (2013) 103526

[18] Y. Kucukakca, Eur.Phys.J. C74 (2014) 10, 3086

[19] M. Jamil, S. Ali, D. Momeni and R. Myrzakulov, Eur.Phys.J. C72 (2012) 1998

[20] S. Basilakos, M. Tsamparlis and A. Paliathanasis, Phys. Rev. D., 83, (2011) 103512

[21] A. Paliathanasis, M. Tsamarlis and S. Basilakos, Phys. Rev. D., 84, (2011) 123514

[22] A. Paliathanasis, M. Tsamparlis, Int. J. Geom. Methods Mod. Phys. 11, 1450037 (2014)

[23] A. Borowiec, S. Capozziello, M. De Laurentis, F. S.N. Lobo, A. Paliathanasis, M. Paolella, A. Wojnar, in press, arXiv:1407.4313 [gr-qc]

[24] M. Tsamparlis, A. Paliathanasis, S. Basilakos and S. Capozziello, Gen. Rel. Grav. **45**, 2003 (2013).

[25] A. Paliathanasis, M. Tsamparlis, S. Basilakos and S. Capozziello, Phys. Rev. D **89**, 063532 (2014).

[26] A. Paliathanasis and M. Tsamparlis, Int. J. Geom. Meth. Mod. Phys. **11**, 1450037 (2014).

[27] B. S. DeWitt, The Many-Worlds Interpretation of Quantum Mechanics, Princeton Series in Physics, Princeton University Press (1973).

[28] S. Capozziello, M. De Laurentis and S. D. Odintsov, Eur. Phys. J. C **72**, 2068 (2012).

Higher-Dimensional Theories with continuous or fuzzy coset spaces as extra dimensions

G. Zoupanos[*], **D. Gavriil**[†], **G. Manolakos**[‡]

Physics Department, National Technical University,
Zografou Campus, GR-15780 Athens, Greece

ABSTRACT

We review the Coset Space Dimensional Reduction (CSDR) scheme and the best model constructed so far. Then we present some details of the CSDR programme, in an alternate version, in which the extra dimensions are considered to be fuzzy. Specifically, we present a four-dimensional $\mathcal{N} = 4$ SYM theory, orbifolded by $\mathbb{Z}_3$, which mimics the behaviour of a dimensionally reduced $\mathcal{N} = 1$, ten-dimensional gauge theory over a set of fuzzy spheres at intermediate high scales. This leads to the trinification GUT $SU(3)^3$ at slightly lower, which in turn can be spontaneously broken to the MSSM in low scales.

1. Introduction

Scientists have set in high priority the aspect of unification of the fundamental forces. Appealing approaches are the ones that support the existence of extra dimensions. A very consistent framework in the unification of all forces supporting such a scenario is Superstring theories [1], with most promising the Heterotic string [2] due to the connections to the low-energy physics. Another remarkable framework for the unification attempt was employed, a few years before the discovery of the Heterotic string, that is the dimensional reduction of higher-dimensional gauge theories. This field was pioneered by Forgacs and Manton with studies on Coset Space Dimensional Reduction (CSDR) [3, 4, 5] and the Scherk-Schwarz group manifold reduction [8]. In these two approaches, a starting gauge theory governs the regime of higher dimensions, where gauge-Higgs unification is achieved, leading to a four-dimensional theory in which the gauge and Higgs fields are the surviving components of the initial fields in high dimensions. In the CSDR scheme, fermions are included in the initial gauge

[*] e-mail address: George.Zoupanos@cern.ch
[†] e-mail address: d.gavriil@yahoo.gr
[‡] e-mail address: gmanol@central.ntua.gr

theory, resulting to Yukawa couplings in four dimensions. The initial theory is required to be $\mathcal{N} = 1$ supersymmetric, i.e. gauge and fermion fields belong to the same vector supermultiplet, relating gauge and fermion fields that have been introduced. Resulting with chiral theories in four dimensions [9, 10] is regarded as a notable achievement.

In order to preserve an $\mathcal{N} = 1$ supersymmetry after the dimensional reduction, Calabi-Yau (CY) spaces are considered as suitable compact internal manifolds [11]. However, the moduli stabilization problem that arose, led to a wider class of internal spaces, called manifolds with SU(3)-structure. Specifically, here we consider an interesting class of SU(3)-structure manifolds, called nearly-Kähler manifolds [12, 14, 15], also see refs from [13].

The homogeneous nearly-Kähler manifolds in six dimensions are the three non-symmetric coset spaces $G_2/SU(3)$, $Sp(4)/(SU(2)\times U(1))_{non-max}$ and $SU(3)/U(1) \times U(1)$ and the group manifold $SU(2) \times SU(2)$ [15] (see also [14, 12]). It is worth noting that in four-dimensional theories resulting from dimensional reduction of a ten-dimensional, $\mathcal{N} = 1$ supersymmetric gauge theory over non-symmetric coset spaces, supersymmetry breaking terms are automatically included [16], [17], contrary to CY spaces.

Another promising framework for describing physics at Planck scale is Non-commutative geometry [18] - [38]. Non-commutative geometry was considered as an appropriate framework for regularizing quantum field theories, or even better, building finite ones. However, constructing quantum field theories on Non-commutative spaces is much more difficult than expected and , furthermore, problematic ultraviolet features have emerged [21] (see also [22] and [23]). Nevertheless, this framework is appropriate to accommodate particle models with Non-commutative gauge theory [24] (see also [25, 26, 27]).

Remarkably, the two frameworks came closer by realizing that in M-theory and "open string theory", in the presence of a non-vanishing background antisymmetric field, the effective physics on D-branes can be described by an Non-commutative gauge theory [28, 29]. Thus, Non-commutative field theories emerge as effective description of string dynamics. Moreover, major contribution in Non-commutative geometry was made by Seiberg and Witten [29]. Their study triggered notable developments [32, 31] and , based on them, Non-commutative versions of SM were built [33]. Unfortunately, those models fail to troubleshoot the main problem of the SM, that is the presence of numerous free parameters, due to the ad hoc consideration of Higgs and Yukawa sectors. Finally, an interesting programme has been suggested and investigated [34, 35, 36, 37, 38] considering the extra dimensions as Non-commutative. This programme overcomes the ultraviolet/infrared problems of theories defined in Non-commutative spaces in an obvious way offering the new possibility to start with an abelian gauge theory defined on the higher-dimensional space and result with a non-abelian one in four dimensions, after dimensional reduction. Additionally, another spectacular feature of this programme is that theories constructed on Non-commutative (fuzzy) manifolds as approximations of the continuous ones, are renormalizable contrary to all known higher-dimensional theories. The latter property was examined from the four-dimensional point of view, too, using spontaneous symmetry breakings, which mimic the results of the dimensional reduction of a higher-dimensional gauge theory with fuzzy extra dimensions. Finally, in this framework, chiral realistic theories have been constructed, too.

2. The Coset Space Dimensional Reduction

The CSDR procedure demands that the field dependence on the extra coordinates is such that the Lagrangian is independent of them. An elegant way to fulfill the above requirement is to allow for a non-trivial dependence on them, in the sense that a symmetry transformation by an element of the isometry group S of the space formed by the extra dimensions B corresponds to a gauge transformation. Along this framework, a gauge invariant Lagrangian will be independent of the extra coordinates. The above mechanism is the basis of the CSDR scheme [3, 4, 5], which assumes that B is a compact coset space, S/R.

In the CSDR scheme one considers a Yang-Mills-Dirac Lagrangian, with gauge group G, defined on a D-dimensional spacetime M^D, with metric g^{MN}, which is compactified to $M^4 \times S/R$, with S/R a coset space. We assume the following form for the metric

$$g^{MN} = \begin{pmatrix} \eta^{\mu\nu} & 0 \\ 0 & -g^{ab} \end{pmatrix}, \tag{1}$$

where $\eta^{\mu\nu} = diag(1,-1,-1,-1)$ and g^{ab} is the coset space metric. The requirement that transformations of the fields under the action of the symmetry group of S/R are compensated by gauge transformations, imposes certain constraints on the fields of our theory. The analysis of these constraints provides us with the four-dimensional unconstrained fields, as well as with the gauge invariance that remains in the theory after dimensional reduction. Therefore, a potential unification of all low energy interactions, gauge, Yukawa and Higgs is achieved.

It is worth noting that the dimensional reduction of higher-dimensional theories results in effective field theories that might contain also towers of massive higher harmonic (Kaluza-Klein) excitations. The behaviour of the running couplings is altered from logarithmic to power [42] by the quantum level contributions of these excitations, resulting in a remarkable change of the traditional unification picture [43]. Using the continuous Wilson renormalization group technique [44], which can be formulated in any number of space-time dimensions, higher-dimensional theories have also been studied at the quantum level, with results in agreement with the treatment involving massive Kaluza-Klein excitations.

2.1. Reduction of a D-dimensional Yang-Mills-Dirac Lagrangian.

Considering a Lie group S and its subgroup R we define a d-dimensional coset S/R on which the extra dimensions of $M^4 \times S/R$ are compactified (M^4 is our space-time). S acts as a symmetry group of the extra coordinates.

According to the CSDR scheme, an S-transformation of the extra d coordinates is a gauge transformation of the fields that are defined on $M^4 \times S/R$, thus a gauge invariant Lagrangian written on this space is independent of the extra coordinates. Fields defined in this way are called symmetric. The d-dimensional gauge field $A_M(x,y)$ is split into its components $A_\mu(x,y)$ and $A_\alpha(x,y)$, corresponding to M^4 and S/R respectively.

Let us now consider a Yang-Mills-Dirac theory with gauge group G defined on a manifold M^D which, as stated, will be compactified to $M^4 \times S/R$, $D = 4 + d$, $d = dimS - dimR$. The action is

$$A = \int d^4x d^d y \sqrt{-g} \left[-\frac{1}{4} Tr(F_{MN} F_{K\Lambda}) g^{MK} g^{N\Lambda} + \frac{i}{2} \bar{\psi} \Gamma^M D_M \psi \right], \quad (2)$$

where $D_M = \partial_M - \theta_M - A_M$, with $\theta_M = \frac{1}{2}\theta_{MN\Lambda}\Sigma^{N\Lambda}$, the spin-connection of M^D, and $F_{MN} = \partial_M A_N - \partial_N A_M - [A_M, A_N]$, where M, N run over the D-dimensional space and A_M and ψ are D-dimensional symmetric fields. The fermion fields can be accommodated in any representation F of G, unless a further symmetry, such as supersymmetry, is required. If we denote by ξ_A^α, $(A = 1, ..., dimS$ and $\alpha = dimR + 1, ..., dimS$ the curved index) the Killing vectors which generate the symmetries of S/R and by W_A the compensating gauge transformation associated with ξ_A, the following constraint equations for scalar ϕ, vector A_α and spinor ψ fields on S/R, are expressing the requirement that transformations of the fields under the action of S/R are compensated by gauge transformations

$$\delta_A \phi = \xi_A^\alpha \partial_\alpha \phi = D(W_A)\phi, \quad (3)$$

$$\delta_A A_\alpha = \xi_A^\beta \partial_\beta A_\alpha + \partial_\alpha \xi_A^\beta A_\beta = \partial_\alpha W_A - [W_A, A_\alpha], \quad (4)$$

$$\delta_A \psi = \xi_A^\alpha \partial_\alpha \psi - \frac{1}{2} G_{Abc} \Sigma^{bc} \psi = D(W_A)\psi, \quad (5)$$

where W_A depend only on internal coordinates y and $D(W_A)$ represents a gauge transformation in the appropriate representation of the fields.

Regarding the constraints (3)-(5), they provide us [3, 6, 4] with the four-dimensional unconstrained fields as well as with the gauge invariance that remains in the theory after dimensional reduction. The components $A_\mu(x,y)$ of the initial gauge field $A_M(x,y)$ become, after dimensional reduction, the four dimensional gauge fields and they are independent of y. Additionally, they have to commute with the elements of the R_G, subgroup of G, meaning that the four-dimensional gauge group H is the centralizer of R in G, $H = C_G(R_G)$. We denote by $\phi_\alpha(x,y)$ the $A_\alpha(x,y)$ components of $A_M(x,y)$. They become scalars in four dimensions and they transform under R as a vector v, i.e.

$$S \supset R \quad (6)$$

$$adjS = adjR + v. \quad (7)$$

Moreover, the $\phi_\alpha(x,y)$ fields act as an interwining operators connecting induced representations of R acting on G and S/R. According to Schur's lemma, the previous expression implies that the transformation properties of the fields $\phi_\alpha(x,y)$ under H can be found, if we decompose the adjoint representation of G according to the embedding:

$$G \supset R_G \times H \tag{8}$$

$$adjG = (adjR, 1) + (1, adjH) + \sum (r_i, h_i). \tag{9}$$

Then, if $v = \sum s_i$, where each s_i is an irreducible representation of R, there survives a Higgs multiplet transforming under the representation h_i of H and all other scalar fields vanish.

Regarding the fermion fields [4, 45, 9, 10] we proceed along similar lines as in the case of scalars. It turns out that the spinor fields act as interwining operators connecting induced representations of R in $SO(d)$ and G. In order to obtain the H representation content of the four-dimensional fermions, we have to decompose the representation F of the initial gauge group, in which the fermions are assigned in higher dimensions, under $R_G \times H$, i.e.

$$F = \sum (t_i, h_i), \tag{10}$$

and the spinor of $SO(d)$ under R

$$\sigma_d = \sum \sigma_j\,. \tag{11}$$

Then for each pair (r_i, σ_i), where r_i and σ_i are identical irreducible representations of R there is an h_i multiplet of spinor fields in the four-dimensional theory. Regarding the possibility of obtaining chiral fermions in the effective theory, we notice that if we start with Dirac fermions in higher dimensions it is impossible to obtain chiral fermions in four dimensions. Further requirements must be considered in order to achieve chiral fermions in the resulting theory. Imposing the Weyl condition in D dimensions, we obtain two sets of Weyl fermions with the same quantum numbers under H. Although this is already a chiral theory, we can go further and try to impose Majorana condition in order to eliminate the doubling of the fermionic spectrum. Majorana and Weyl conditions are compatible in $D = 4n + 2$ which is the case of our interest.

The allowed embeddings of R into G are restricted by the condition that an anomaly free theory in higher dimensions must fulfill, in order to obtain anomaly free theories in four dimensions after the dimensional reduction [46]. According to that condition, the allowed embeddings are related with the embedding of R into $SO(6)$, the tangent space of the six-dimensional cosets we consider [4, 40, 7]. According to ref. [7] the anomaly cancelation condition is automatically satisfied for the choice of embedding $E_8 \supset SO(6) \supset R$, which we adopt here.

2.2. Dimensional Reduction of E_8 over $SU(3)/U(1) \times U(1)$

In this subsection we summarize a few results concerning the dimensional reduction of the $\mathcal{N} = 1, E_8$ SYM over $SU(3)/U(1) \times U(1)$ [39]. The four-dimensional gauge group will be provided by the decomposition of E_8 under $R = U(1) \times U(1)$ suggested by

$$E_8 \supset E_6 \times SU(3) \supset E_6 \times U(1)_A \times U(1)_B \,. \tag{12}$$

According to the rules of the previous section, the surviving gauge group in four dimensions is

$$H = C_{E_8}(U(1)_A \times U(1)_B) = E_6 \times U(1)_A \times U(1)_B \,. \tag{13}$$

The surviving scalars and fermions in four dimensions are provided by the explicit decomposition of the adjoint representation of E_8, 248 under $U(1)_A \times U(1)_B$. Applying the CSDR rules we find that the resulting four-dimensional theory is an $\mathcal{N} = 1$, E_6 GUT with $U(1)_A, U(1)_B$ as global symmetries. The potential is determined by the decomposition of the specific $S = SU(3)$ under $R = U(1) \times U(1)$ studied in [17]. The D-terms can be constructed and the F-terms are obtained by the superpotential. The rest of the terms in the potential could be interpreted as soft scalar masses and trilinear soft terms. Finally, the gaugino obtains a mass and receives contribution from the torsion contrary to the rest soft supersymmetry breaking terms.

2.3. $SU(3)^3$ due to Wilson flux

According to the previous section, the $E_6 \times U(1) \times U(1)$ group is the surviving gauge group of the initial's E_8 group dimensional reduction. The surviving scalars in the four-dimensional theory, being in the fundamental representation of the gauge group are not able to provide the appropriate symmetry breaking towards the standard model. In order to reduce further the gauge symmetry, one has to apply the Wilson flux breaking mechanism [49, 48, 47]. Application of this mechanism imposes further constraints in the scheme.

In the case of our interest, instead of considering the simply connected manifold B_0, where B_0 is the coset S/R, we consider the multiply connected manifold $B = B_0/F^{S/R}$ with $F^{S/R}$ a freely acting discrete symmetry of B_0. The manifold B is multiply connected due to the presence of the symmetry $F^{S/R}$. For each element $g \in F^{S/R}$, we pick up an element U_g in H, which can be represented as the Wilson loop. If the manifold is simply connected, then the vanishing of the field strength ensures that we can set the gauge field to zero by a gauge transformation. In the case of a multiply connected manifold, although the vacuum field strength vanishes everywhere, U_g cannot be set to one and the gauge field cannot be set to zero. Therefore, a homomorphism of $F^{S/R}$ into H is induced with image T^H, which is the subgroup of H generated by the element U_g.

Concerning the gauge symmetry that is preserved by the vacuum, we consider the following. The vacuum has $A^a_\mu = 0$ and we represent a gauge transformation by a space-dependent matrix $V(x)$ of H. In order to keep $A^a_\mu = 0$ and leave the vacuum invariant, $V(x)$ must be a constant. Moreover, the matrix $V(x)$ is consistent with the action of the elements U_g only if $[V, U_g] = 0$ for all $g \in F^{S/R}$. Therefore, the unbroken subgroup of H is the centralizer of T^H in H. Respectively, the surviving matter fields are those that are invariant under the diagonal sum $F^{S/R} \oplus T^H$. The discrete symmetries $F^{S/R}$, which act freely on coset spaces $B_0 = S/R$ are the center of S, $Z(S)$ and $W = W_S/W_R$, where W_S and W_R are the Weyl groups of S and R, respectively. In the case of our interest, where $B_0 = SU(3)/U(1) \times U(1)$, we have $F^{S/R} = \mathbb{Z}_3 \subseteq W$. After the $\mathbb{Z}_3$ projection, the gauge group E_6 breaks to $SU(3)_C \times SU(3)_L \times SU(3)_R$, (the first of the $SU(3)$ factors is the Standard Model colour gauge group). Moreover, one can obtain three fermion generations by introducing non-trivial monopole charges in the $U(1)$'s in R.

In ref [14] it was shown that the scalar potential leads to the proper hierarchy of spontaneous breaking. Using the appropriate vev's, a first spontaneous symmetry breaking leads to the MSSM [50], while the electroweak breaking proceeds by a second one [51]. It is worth noting that before the EW symmetry breaking, supersymmetry is broken by both D-terms and F-terms, in addition to its breaking by the soft terms.

We plan to examine in detail the phenomenological consequences of the resulting model, taking also into account the massive Kaluza-Klein modes.

3. Fuzzy spaces and fuzzy dimensional reduction

In order to continue our analysis, it is fundamental to introduce the concept of the fuzzy sphere [41]. The appropriate way to do so, is to initially consider the ordinary sphere S^2, on which the algebra of functions is commutative, and then define the fuzzy sphere as its extension.

It is known that the algebra of functions on S^2 is generated by the spherical harmonics, Y_{lm}, i.e. any arbitrary function on S^2 can be expanded in terms of Y_{lm}, since they form a complete and orthogonal set of functions. In the fuzzy case (the most typical case of Non-commutative geometry), contrary to the non-fuzzy sphere, the integer number l does have an upper limit. So, the algebra of functions on the fuzzy sphere is truncated to finite dimensional - naturally considered as a matrix algebra. Therefore, it proves that it is consistent to define the fuzzy sphere[1] as a matrix approximation of the non-fuzzy sphere and that the truncation of the algebra of the functions[2] is responsible for the deprivation of commutativity.

Given that we aim at studying gauge theory on fuzzy sphere, the next -obvious- step is to examine the behaviour of the gauge fields on the fuzzy sphere. So, we consider a field $\phi(X_a)$ on the fuzzy sphere, with X_a being

[1]The geometry of other (higher-dimensional) fuzzy spaces (e.g. fuzzy CP^M) are examined in [52, 53]

[2]Besides of functions on the fuzzy sphere, spinors can be examined as well [34]

the covariant coordinates [54] and then we take an infinitesimal transformation of this field

$$\delta\phi(X) = \lambda(X)\phi(X)\,, \tag{14}$$

where $\lambda(X)$ is the parameter of the gauge transformation. Under the above gauge transformation it holds that $\delta X_a = 0$, ensuring the invariance of the covariant coordinates. Therefore, in the Non-commutative case, when left multiplying by a coordinate, we obtain

$$\delta(X_a\phi) = X_a\lambda(X)\phi\,, \tag{15}$$

and in general, it holds that $X_a\lambda(X)\phi \neq \lambda(X)X_a\phi$. So, according to the non-fuzzy gauge theory, one needs to introduce the covariant coordinates phi_a, in order to obtain $\delta(\phi_a\phi) = \lambda\phi_a\phi$, with $\delta(\phi_a) = [\lambda, \phi_a]$. Also, it is set that $\phi_a \equiv X_a + A_a$, where A_a is the gauge potential, concluding in the equivalence that ϕ_a is the analogue of the covariant derivative of the original gauge theories. From the above equations, the transformation of A_a is

$$\delta A_a = -[X_a, \lambda] + [\lambda, A_a]\,, \tag{16}$$

encouraging the identification of A_a with a gauge field.

4. Fuzzy CSDR

Attempts to reproduce the dominant gauge theory that describes physics in low energies are based on the above structure. More specifically, we consider a Non-commutative gauge theory on the $M^4 \times (S/R)_F$[3] space, then we perform dimensional reduction and in the end we result with a four dimensional theory. Unfortunately, realistic results did not arise using this method, therefore, in order to obtain a more appropriate gauge theory in four dimensions, a non trivial dimensional reduction had to be applied, namely the fuzzy extension of the CSDR scheme.

The factor that differentiates the fuzzy CSDR from the original one, is the consideration of the extra dimensions as fuzzy coset spaces [34] (see also [55]), meaning that the group S acts now on the fuzzy coset $(S/R)_F$, with the fields remaining invariant under an infinitesimal transformation of S - up to an infinitesimal gauge transformation. Specifically, the fuzzy coset we make use is the fuzzy sphere, $(SU(2)/U(1))_F$, therefore scalar and gauge fields should be left invariant under an infinitesimal transformation of $SU(2)$ on the fuzzy sphere, up to an infinitesimal gauge transformation

$$\mathcal{L}_b\phi = \delta_{W_b} = W_b\phi \tag{17}$$

$$\mathcal{L}_b A = \delta_{W_b} A = -DW_b\,, \tag{18}$$

where A is the gauge potential and W_b is an antihermitian gauge parameter which depends on X_a. Therefore, W_b can be written as

$$W_b = W_b^I \mathcal{T}^I\,, \quad I = 1, 2, \ldots, P^2\,, \tag{19}$$

[3] $(S/R)_F$ is a fuzzy coset e.g. the fuzzy sphere.

where $\mathcal{T}^I$ are the hermitian generators of the gauge group of the theory $U(P)$ and $(W_b^I)^\dagger = -W_b^I$. The CSDR constraints are converted in the form

$$[\omega_b, A_\mu] = 0 \tag{20}$$
$$C_{bde}\phi^e = [\omega_b, \phi_d]\,, \tag{21}$$

where $\phi_a \equiv X_a + A_a$ -as mentioned above- and $\omega_a \equiv X_a - W_a$. Since Lie derivatives respect the $\mathfrak{su}(2)$ commutation relations, one is led to the consistency condition

$$[\omega_a, \omega_b] = C_{ab}^c \omega_c\,, \tag{22}$$

where ω_a transforms as $\omega_a \to \omega_a' = g\omega_a g^{-1}$. As for the spinors, a quite similar procedure that is followed [34].

As an application of the fuzzy CSDR scheme, we present the example, where the gauge group is $U(1)$ and the fuzzy coset is the fuzzy sphere. The ω_a are $N \times N$ antihermitian matrices, therefore they can be considered as elements of $U(N)$. Though, the consistency relation, (22), must hold, that is the ω_a obeys the commutation relation of the $SU(2)$ algebra. Thus, the $SU(2)$ algebra has to be embedded into the $U(1)$ algebra. So, let $T^h, h = 1\ldots, N^2$ be the generators of the $U(N)$ in the fundamental representation, and make use of the convention $h(a,u), a = 1,2,3, u = 4,5,\ldots,N^2$, with the generators T^a satisfying the $SU(2)$ algebra $[T^a, T^b] = C_c^{ab}T^c$. Obviously the embedding is achieved with the identification $\omega_a = T_a$.

Let us now examine and give interpretations of the two constraints (20), (21). The first one suggests that the gauge group of the four-dimensional theory is the centralizer of the image of $SU(2)$ in $U(N)$, that is

$$K = C_{U(N)}(SU(2)) = SU(N-2) \times U(1) \times U(1)\,. \tag{23}$$

Therefore, there is an arbitrariness on the dependance of $A_\mu(x, X)$ on x, but as for X, they depend on them meaning that the latter are valued in K instead of $U(N)$. Rephrasing, the 4-dimensional gauge potential that one is led is valued in K. The second constraint is satisfied after choosing $\phi_a = r\phi(x)\omega_a$. This means that the remaining unconstrained degrees of freedom are related to the scalar field $\phi(x)$, which belongs to the trivial representation of the 4-dimensional gauge group K.

Summing up the above procedure, one starts with a gauge theory which is described by a $U(1)$ on $M^4 \times S_N^2$. The consistency condition is satisfied by embedding the $SU(2)$ into $U(N)$[4]. Then, imposing the two CSDR constraints, the four-dimensional group is obtained and the scalar fields that do survive the reduction procedure arise.

Let us now proceed with listing the results of the above procedure, for the fermionic case. The extended analysis [34] proves that the appropriate embedding is $S \subset SO(\text{dim}S)$, which is achieved by $T_a = \frac{1}{2}C_{abc}\Gamma^{bc}$, which respects the $SU(2)$ commutation relation. Therefore, ψ is an interwining

[4] Instead of embedding the $SU(2)$ into the fundamental representation of $U(N)$, one could have used other representations, too [41]

operator between the representations of S and $SO(\text{dim}S)$. According to the commutative case [4], the surviving fermions in four-dimensional theory arise by decomposing the adjoint representation of $U(NP)$ under the product $S_{U(NP)} \times K$, that is

$$U(NP) \supset S_{U(NP)} \times K\,, \quad \text{adj}\, U(NP) = \sum_i (s_i, k_i)\,. \tag{24}$$

Moreover, the decomposition of the spinorial representation σ of $SO(\text{dim}S)$ under S is

$$SO(\text{dim}S) \supset S\,, \quad \sigma = \sum_e \sigma_e\,. \tag{25}$$

Therefore, in case that the two irreducible representations s_i, σ_e are identical, the fermions that survive (4-dimensional spinors) and are present in the four-dimensional theory, belong to the k_i representation of K.

Ending this section, it is important to compare the ordinary higher-dimensional theory $M^4 \times (S/R)$, to the fuzzy one, $M^4 \times (S/R)_F$. Both theories have the same isometries - fuzziness does not affect them -, i.e. $SO(1,3) \times SO(3)$. In addition, the dimensionality of the gauge couplings defined on the two spaces is the same. On the other side, they present a very striking difference: Non-commutative higher-dimensional theory is the only one that is renormalizable[5]. Moreover, a $U(1)$ gauge group defined on the $M^4 \times (S/R)_F$ space, is appropriate in order to end up with a non-abelian four-dimensional theory [6].

5. Orbifolds and fuzzy extra dimensions

The recovering of chiral four-dimensional theories starting from higher-dimensional theories with fuzzy extra dimensions was the motivation of the introduction of the orbifold structure, similar to the one in [56]. The orbifold procedure offers an alternative way to obtain $\mathcal{N} = 1$ four-dimensional models after reducing a higher-dimensional theory on appropriate manifolds, e.g. Calabi-Yau [57] or $SU(3)$ structured ones [12, 58]. Duality between four-dimensional $\mathcal{N} = 4$, $U(N)$ SYM theory and Type IIB string theory on $AdS_5 \times S^5$ [59], motivated the authors of [56] to proceed to the application of orbifold techniques - similar to [60, 61] - in order to break some of the four supersymmetries. Moreover, the initial gauge group, $SU(3N)$, that is realized on $3N$ $D3$ branes[7], breaks to $SU(N)^3$ with fermions being accommodated into its chiral representations.

The concept of deconstructing dimensions [62], motivated the idea to reverse the above procedure [35, 36, 37] for further justification of the

[5]Meaning that the divergencies are eliminated by a finite number of counter-terms.

[6]Technically, this is possible because $N \times N$ matrices could be decomposed on the $U(N)$ generators.

[7]This is point where the two different frameworks (superstring theories and Non-commutative geometry) that aim at unification meet, i.e. Non-commutative gauge theory can describe effective physics on D-branes.

renormalizability of the theory and construction of chiral models in theories arising from the framework of fuzzy extra dimensions. Reversing the procedure gives hope that consideration of the initial abelian gauge theory as a higher-dimensional one is not necessary, instead the non-abelian gauge theory could emerge from fluctuations of the coordinates [63]. Realizing the last consideration, one has to start with a four-dimensional gauge theory, including an appropriate scalar spectrum and a suitable potential producing vacua that could be interpreted as dynamically generated fuzzy extra dimensions including, at the same time, a finite Kaluza-Klein tower of massive modes. Also, although in such models the inclusion of chiral fermions is preferred, the best one achieved so far includes mirror fermions[8] [36, 37].

In this review, the above sketch is realized performing a dimensional reduction on an orbifold [65, 64]. More specifically, we examine the spectrum of the surviving fields and the superpotential of the projected theory, after the application of $\mathbb{Z}_3$ orbifold projection of the $\mathcal{N}=4$ SYM theory [66]. In our case, this theory is an $SU(3N)$ and the particle content is one $SU(3N)$ gauge supermultiplet and three adjoint chiral supermultiplets. Their component fields are the gauge bosons, six adjoint real scalars and four adjoint Weyl fermions. The scalars and Weyl fermions transform under representations of the $SU(4)_R$ symmetry of the theory - 6, 4 respectively. That is the reason why - in order to introduce orbifolds - the discrete group $\mathbb{Z}_3$ must be included as a subgroup of $SU(4)_R$. Although there are more than one options, the appropriate one is to embed the discrete group into the $SU(3)$ subgroup of $SU(4)_R$. The suitability of this choice lies into the fact that it is the only one that leads to the desired $\mathcal{N}=1$ supersymmetric models [56] (with $U(1)_R$ R-symmetry). Since the particles that consist the spectrum of the theory belong to different representations of $SU(4)_R$, it is expected that $\mathbb{Z}_3$ will act non-trivially on them. In the case of gauge and gaugino fields, the action of $\mathbb{Z}_3$ is trivial, since they are singlets under $SU(4)_R$. On the other hand, scalars and fermions will transform non trivially under the above action. Specifically, as far as the matter fields is concerned, since they are not invariant under a gauge transformation, $\mathbb{Z}_3$ acts on their gauge indices, too. Therefore, the orbifold filters this way the particle spectrum and the derived theory contains the particles which are invariant under the combined $\mathbb{Z}_3$ action on both the geometric[9] and gauge indices [61].

So - after the orbifold projection - the gauge group of the initial theory breaks down to the group $H = SU(N) \times SU(N) \times SU(N)$ with both scalar and fermionic fields transforming under the same representation $3 \cdot \big((N, \bar{N}, 1) + (\bar{N}, 1, N) + (1, N, \bar{N})\big)$, a result that demonstrates the presence of the $\mathcal{N}=1$ remnant supersymmetry. The chiral supermultiplet, which fermions and scalars share, is an anomaly free representation of H.

[8]Ending up with mirror fermions does not forbid phenomenological contact [79], however exactly chiral fermions are preferred.

[9]In case of ordinary reduction of a 10-dimensional $\mathcal{N}=1$ SYM theory, one obtains an $\mathcal{N}=4$ SYM theory in four dimensions with a global $SU(4)_R$ symmetry which is identified with the tangent space $SO(6)$ of the extra dimensions [16, 17].

Besides the particle spectrum of the projected theory, it is necessary to determine its superpotential, which is derived from the superpotential of the initial $\mathcal{N}=4$ SYM theory [66]

$$W_{\mathcal{N}=4} = \epsilon_{ijk} \mathrm{Tr}(\Phi^i \Phi^j \Phi^k) \,, \tag{26}$$

where the Φ's are chiral superfields. The above structure remains the same after the projection, but it encodes only the surviving fields of the $\mathcal{N}=1$ theory that passed the orbifold filtering

$$W^{(proj)}_{\mathcal{N}=1} = \sum_I \epsilon_{ijk} \Phi^i_{I,I+a_i} \Phi^j_{I+a_i,I+a_i+a_j} \Phi^k_{I+a_i+a_j,I} \,. \tag{27}$$

The next step is to find the scalar potential of the projected thepry. This can be achieved by extracting it from the above superpotential, $W^{(proj)}_{\mathcal{N}=1}$. Therefore, the scalar potential is

$$V^{proj}_{\mathcal{N}=1}(\phi) = \frac{1}{4} \mathrm{Tr}\left([\phi^i, \phi^j]^\dagger [\phi^i, \phi^j] \right) \,, \tag{28}$$

where, ϕ^i are the scalar component fields of the chiral superfield Φ^i. Unfortunately, the minimization of $V^{proj}_{\mathcal{N}=1}(\phi)$ is achieved only by vanishing vevs of the fields, therefore, it is necessary to modify it in order that solutions which could be interpreted as vacua of a Non-commutative geometry to be emerged. So, addressed to this direction, $\mathcal{N}=1$ soft supersymmetric terms of the form[10]

$$V_{SSB} = \frac{1}{2} \sum_i m_i^2 \phi^{i\dagger} \phi^i + \frac{1}{2} \sum_{i,j,k} h_{ijk} \phi^i \phi^j \phi^k + h.c. \tag{29}$$

are inserted into $V^{proj}_{\mathcal{N}=1}(\phi)$, where $h_{ijk}=0$, unless $i+j+k=0 \mathrm{mod} 3$. It is important to refer that the introduction of the SSB terms does not cause embarrassment, since an SSB sector is indispensable for a model aspired to be realistic, see e.g. [68].

So, the total scalar potential is

$$V = V^{proj}_{\mathcal{N}=1} + V_{SSB} + V_D \,, \tag{30}$$

where the term V_D represents the D-terms of the theory, which are given by

$$V_D = \frac{1}{2} D^2 = \frac{1}{2} D^I D_I \,, \tag{31}$$

[10]Only purely scalar SSB terms will be inserted into the scalar potential. Of course, other SSB terms have to be included in order to obtain the full SSB sector [68], however it is not necessary for our purposes.

where $D^I = \phi_i^\dagger T^I \phi^i$ and T^I are the generators of the chiral multiplets - in the representation they belong. Fixing the parameters of the (29) to $m_i^2 = 1\,,\quad h_{ijk} = \epsilon_{ijk}$, the total scalar potential turns to be

$$V = \frac{1}{4}(F^{ij})^\dagger F^{ij} + V_D\,, \tag{32}$$

where F^{ij} is[11]

$$F^{ij} = [\phi^i, \phi^j] - i\epsilon^{ijk}(\phi^k)^\dagger\,. \tag{33}$$

Obviously, the first term of (32) is positive definite, which means that the global minimum of the potential is obtained if

$$[\phi^i, \phi^j] = i\epsilon_{ijk}(\phi^k)^\dagger\,,\quad \phi^i(\phi^j)^\dagger = R^2\,, \tag{34}$$

with $[R^2, \phi^i] = 0$. It seems that the fuzzy sphere underlies in the above equations, so it just remains to designate it. This will arise by defining the untwisted fields $\tilde{\phi}^i$ as $\phi^i = \Omega\tilde{\phi}^i$, with $\Omega \neq 1$, satisfying the following relations

$$\Omega^3 = 1\,,\quad [\Omega, \phi^i] = 0\,,\quad \Omega^\dagger = \Omega^{-1}\,,\quad (\tilde{\phi}^i)^\dagger = \tilde{\phi}^i \;\Leftrightarrow\; (\phi^i)^\dagger = \Omega\phi^i\,. \tag{35}$$

Now, it is clear that (34) reduce to the ordinary fuzzy sphere relation generated by $\tilde{\phi}^i$

$$[\tilde{\phi}^i, \tilde{\phi}^j] = i\epsilon_{ijk}\tilde{\phi}^k\,,\quad \tilde{\phi}^i\tilde{\phi}^i = R^2\,. \tag{36}$$

This demonstrates the reason why the Non-commutative space that generates the ϕ^i is called twisted fuzzy sphere, $\tilde{S}_N^2$. The fact that the above structure is valid only for $\mathbb{Z}_3$, poses it as the unique choice as the appropriate cyclic group for the orbifold projection.

A configuration of the twisted fields, ϕ^i, that satisfy (34) is $\phi^i = \Omega(\mathbf{1}_3 \otimes \lambda^i_{(N)})$, where $\lambda^i_{(N)}$ are the generators in the N-dimensional irreducible representation of $SU(2)$ and Ω is the matrix described by the following relations:

$$\Omega = \Omega_3 \otimes \mathbf{1}_N\,,\quad \Omega_3 = \begin{pmatrix} 0 & 1 & 0 \\ 0 & 0 & 1 \\ 1 & 0 & 0 \end{pmatrix}\,,\quad \Omega^3 = \mathbf{1}\,. \tag{37}$$

According to the transformation $\phi^i = \Omega\tilde{\phi}^i$, the "off-diagonal" orbifold sectors are expressed in block-diagonal form

$$\phi^i = \begin{pmatrix} 0 & (\lambda^i_{(N)})_{(N,\bar{N},1)} & 0 \\ 0 & 0 & (\lambda^i_{(N)})_{(1,N,\bar{N})} \\ (\lambda^i_{(N)})_{(\bar{N},1,N)} & 0 & 0 \end{pmatrix} = \Omega \begin{pmatrix} \lambda^i_{(N)} & 0 & 0 \\ 0 & \lambda^i_{(N)} & 0 \\ 0 & 0 & \lambda^i_{(N)} \end{pmatrix}. \tag{38}$$

[11] The V_D term settles for a change on the radius of the sphere, in accordance to the ordinary fuzzy sphere case [35, 69, 37]

It is clear that the (untwisted) fields, $\tilde{\phi}^i$, that generate the ordinary fuzzy sphere, have taken a block-diagonal form. Each block separately can be regarded as an ordinary fuzzy sphere, since the corresponding commutation relations (36) they separately satisfied. Therefore, the configuration (38) could be interpreted as three separate fuzzy spheres (branes), embedded with relative angles $2\pi/3$. Rephrasing, the solution ϕ^i is equivalent to three fuzzy spheres which conform with the orbifolding. In a few words, the global minimum of the scalar potential - at least for a fixed range of parameters - is achieved by a twisted fuzzy sphere. So, the F^{ij} that was defined in (33), could be considered as the field strength on the spontaneously generated fuzzy extra dimensions.

Let us now examine the potential's vacuum from a geometric point of view. Fixing the parameters, the potential gets minimized by a twisted fuzzy sphere solution

$$\phi^i = \Omega(\mathbf{1}_3 \otimes (\lambda^i_{(N-n)}) \oplus 0_n)\,, \tag{39}$$

where 0_n is the $n \times n$ matrix with zero entries. This non-vanishing vacuum - a vacuum considered as $\mathbb{R}^4 \times \tilde{S}^2_N$ with a twisted fuzzy sphere in the ϕ^i coordinates - breaks the gauge symmetry, $SU(N)^3$ down to $SU(n)^3$.

The fluctuations of the scalar fields around the vacuum can be understood by considering the transformation, $\phi^i = \Omega\tilde{\phi}^i$. In the non-twisted case, fluctuations around the vacuum describe scalar and gauge fields on S^2_N [54, 63], which gain mass from the $\mathbb{R}^4$ point of view. The (38) exhibits that the matrix Ω maps the twisted fuzzy sphere into three ordinary fuzzy spheres as three $N \times N$ blocks are diagonally embedded into the original $3N \times 3N$ matrix. Therefore, all fluctuations could be considered as fields on the three untwisted fuzzy spheres

$$\phi^i = \tilde{\Omega}(\lambda^i_{(N)} + A^i) = \begin{pmatrix} \lambda^i_{(N)} + A^i & 0 & 0 \\ 0 & \omega(\lambda^i_{(N)} + A^i) & 0 \\ 0 & 0 & \omega^2(\lambda^i_{(N)} + A^i) \end{pmatrix}, \tag{40}$$

as well as specific massive off-diagonal states which cyclically connect these spheres. The field strength F^{ij}, (33), converts to

$$F^{ij} = [\phi^i, \phi^j] - i\epsilon^{ijk}(\phi^k)^\dagger = \Omega^2\left([\tilde{\phi}^i, \tilde{\phi}^j] - i\epsilon^{ijk}\tilde{\phi}^k\right), \tag{41}$$

that is the field strength on an untwisted fuzzy sphere. Thus, at intermediate energy scales, the vacuum can be interpreted as $\mathbb{R}^4 \times S^2_N$ with three untwisted fuzzy spheres in the $\tilde{\phi}^i$ coordinates. The three spheres do not mix, due to the lack of off-diagonal entries, due to the orbifold projection. As in [35, 36, 37], because of the Higgs effect, fermions and gauge fields decompose to a finite Kaluza-Klein tower of massive modes on S^2_N resp. $\tilde{S}^2_N$, as well as a massless sector.

6. The $SU(3)_c \times SU(3)_L \times SU(3)_R$ chiral model

Working in the above context, three (minimal and anomaly free) models emerge, however we will focus on the most interesting one. For all models the initial theory is the same, i.e. a $\mathcal{N}=1$ SYM 4-dimensional $SU(3N)$ gauge theory whose field spectrum was listed in the previous section. Then it follows the realization of the orbifold projection of the theory, embedding - as we have already noted - the $\mathbb{Z}_3$ into the $SU(3)$ subgroup of $SU(4)_R$. After the projection, the initial gauge group breaks to the $\mathcal{N}=1$ $SU(N)^3$ and the fields of the theory are accommodated into chiral representations of the gauge group. More specifically, there are three families transforming as

$$(N,\bar{N},1)+(\bar{N},1,N)+(1,N,\bar{N}) \tag{42}$$

under the gauge group $SU(N)^3$. Of course, due to the different ways the initial gauge group $SU(3N)$ is spontaneously broken, we end up with different unification groups $SU(4)\times SU(2)\times SU(2)$, $SU(4)^3$ and $SU(3)^3$[12].

Let us now focus on the very interesting trinification group $SU(3)_c \times SU(3)_L \times SU(3)_R$ [70, 71] (see also [72, 73, 74, 75, 76]; for a string theory approach see [77]). At first, we need to decompose the integer N as $N = n+3$ and then - for the $SU(N)$ - we consider the embedding

$$SU(N) \supset SU(n)\times SU(3)\times U(1)\,, \tag{43}$$

from which it is follows that the embedding for the total gauge group $SU(N)^3$ is

$$SU(N)^3 \supset SU(n)\times SU(3)\times SU(n)\times SU(3)\times SU(n)\times SU(3)\times U(1)^3\,. \tag{44}$$

The three $U(1)$ factors[13] are ignored and according to the above decomposition, the representations (42) decompose (44), (after reordering the factors) as

$$\begin{aligned}
&SU(n)\times SU(n)\times SU(n)\times SU(3)\times SU(3)\times SU(3)\,,\\
&(n,\bar{n},1;1,1,1)+(1,n,\bar{n};1,1,1)+(\bar{n},1,n;1,1,1)+(1,1,1;3,\bar{3},1)\\
&+(1,1,1;1,3,\bar{3})+(1,1,1;\bar{3},1,3)+(n,1,1;1,\bar{3},1)+(1,n,1;1,1,\bar{3})\\
&+(1,1,n;\bar{3},1,1)+(\bar{n},1,1;1,1,3)+(1,\bar{n},1;3,1,1)+(1,1,\bar{n};1,3,1)\,.
\end{aligned} \tag{45}$$

So, judging from the decomposition (43), the gauge group is broken to $SU(3)^3$. The surviving fields transform under the gauge group $SU(3)^3$, as

$$SU(3)\times SU(3)\times SU(3)\,, \tag{46}$$
$$((3,\bar{3},1)+(\bar{3},1,3)+(1,3,\bar{3}))\,, \tag{47}$$

[12] Similar approaches have been examined in the YM matrix models framework [78], but they deprived of phenomenological viability.

[13] These factors decouple at the low energy sector of the theory due to a possible gaining of mass by the Green-Schwarz mechanism [67].

which correspond to the desired chiral representations of the trinification group. The transformation of quarks and leptons (only for the first family but it is similar for the other two) under the gauge group $SU(3)_c \times SU(3)_L \times SU(3)_R$ is

$$q = \begin{pmatrix} d & u & h \\ d & u & h \\ d & u & h \end{pmatrix} \sim (3, \bar{3}, 1) \,, \quad q^c = \begin{pmatrix} d^c & d^c & d^c \\ u^c & u^c & u^c \\ h^c & h^c & h^c \end{pmatrix} \sim (\bar{3}, 1, 3) \,,$$

$$\lambda = \begin{pmatrix} N & E^c & \nu \\ E & N^c & e \\ \nu^c & e^c & S \end{pmatrix} \sim (1, 3, \bar{3}) \,, \tag{48}$$

respectively.

It is crucial to mention that this theory could be upgraded to a two-loop finite theory (see reviews [80, 81, 82, 51]) and furthermore could make testable predictions [51].

Moreover, the fuzzy orbifold mechanism can be used to break spontaneously the unification gauge group down to MSSM and then, in turn, to the $SU(3)_c \times U(1)_{em}$. Concluding, it is useful to sum up the general idea of the theoretical model. At the very high-scale regime, there is an unbroken renormalizable gauge theory. After the spontaneous symmetry breaking, the resulting gauge theory we are led to, is an $SU(3)^3$ GUT, accompanied by a finite tower of massive Kaluza-Klein modes. Finally, in the low scale regime, the trinification group $SU(3)^3$ breaks down to the MSSM. Thus, we conclude with the statement that fuzzy extra dimensions can be used to construct chiral, renormalizable and phenomenologically viable field-theoretical models.

A natural extension of the above ideas and methods have been reported in ref [83] (see also [84]), realized in the context of Matrix Models (MM). At a fundamental level, the MMs introduced by Banks-Fischler-Shenker-Susskind (BFSS) and Ishibashi-Kawai-Kitazawa-Tsuchiya (IKKT), are supposed to provide a non-perturbative definition of M-theory and type IIB string theory respectively [30, 85]. On the other hand, MMs are also useful laboratories for the study of structures which could be relevant from a low-energy point of view. Indeed, they generate a plethora of interesting solutions, corresponding to strings, D-branes and their interactions [30, 86], as well as to non-commutative/fuzzy spaces, such as fuzzy tori and spheres [87]. Such backgrounds naturally give rise to non-abelian gauge theories. Therefore, it appears natural to pose the question whether it is possible to construct phenomenologically interesting particle physics models in this framework as well. In addition, an orbifold MM was proposed by Aoki-Iso-Suyama (AIS) in [88] as a particular projection of the IKKT model, and it is directly related to the construction described above in which fuzzy extra dimensions arise with trinification gauge theory [38]. By $\mathbb{Z}_3$ - orbifolding, the original symmetry of the IKKT matrix model with matrix size $3N \times 3N$ is generally reduced from $SO(9,1) \times U(3N)$ to $SO(3,1) \times U(N)^3$. This model is chiral and has $D = 4$, $\mathcal{N} = 1$ supersymmetry of Yang-Mills type as well as an inhomogeneous supersymmetry specific to matrix models. The $\mathbb{Z}_3$ - invariant fermion fields transform as bi-fundamental representations under the unbroken gauge symmetry exactly as in the constructions described above. In the future we plan to extend further the studies initiated in refs [83, 84] in the context of orbifolded IKKT models.

Acknowledgement: This research is implemented under the Research Funding Program ARISTEIA, Higher Order Calculations and Tools for High Energy Colliders, HOCTools and the ARISTEIA II, Investigation of certain higher derivative term field theories and gravity models (co-financed by the European Union (European Social Fund ESF) and Greek national funds through the Operational Program Education and Lifelong Learning of the National Strategic Reference Framework (NSRF)), as well as the European Union's ITN programme HIGGSTOOLS.

One of us, G.Z, would like to thank the organizers for the warm hospitality.

References

[1] M.B. Green, J.H. Schwarz, E. Witten, Cambridge Monographs on Mathematical Physics, Cambridge University Press, Cambridge, 1987; Polchinski J., Cambridge University Press, Cambridge, 1998; Blumenhagen R., Lüst D., Theisen S., Basic concepts of string theory, Springer, 2013.

[2] D.J. Gross et al., Nucl. Phys. B256 (1985) 253; ibid, Phys. Rev. Lett. 54 (1985) 502.

[3] P. Forgács, N.S. Manton, Comm. Math. Phys. 72 (1980) 15-35.

[4] D. Kapetanakis, G. Zoupanos, Phys. Rep. 219 (1992).

[5] Yu.A. Kubyshin et al., Lecture Notes in Physics, Vol. 349, Springer-Verlag, Berlin, 1989.

[6] E. Witten, Phys. Rev. Lett. 38, 121 (1977).

[7] K. Pilch and A. N. Schellekens, Nucl. Phys. B259, 673(1985); D. Luest, Nucl. Phys. B276, 220(1985)

[8] J. Scherk, J.H. Schwarz, Nucl. Phys. B153 (1979) 61-88.

[9] N.S. Manton, Nucl. Phys. B193 (1981) 502-516.

[10] G. Chapline, R. Slansky, Nucl. Phys. B209 (1982) 461-483.

[11] P. Candelas et al., Nucl. Phys. B258, (1985) 46

[12] G.L. Cardoso et al., Nucl. Phys. B 652 (2003) 5-34, hep-th/0211118; A. Strominger, Nucl. Phys. B274 (1986) 253; D. Lüst, Nucl.Phys. B276 (1986) 220; Castellani L., Lüst D., Nucl.Phys. B296 (1988) 143.

[13] D. Gavriil, G. Manolakos, G. Zoupanos, [arXiv:1412.0438[hep-th]].

[14] N. Irges and G. Zoupanos, Phys. Lett. B698, (2011) 146 [arXiv:hep-ph/1102.2220]; N. Irges, G. Orfanidis, G. Zoupanos, [arXiv:1205.0753 [hep-ph]], PoS CORFU2011 (2011) 105.

[15] J. B. Butruille, [arXiv:math.DG/0612655].

[16] P. Manousselis, G. Zoupanos, Jun 2001, Phys.Lett. B518 (2001) 171-180, hep-ph/0106033; ibid, Phys.Lett. B504 (2001) 122-130

[17] P. Manousselis, G. Zoupanos, Jun 2004, JHEP 0411 (2004) 025, hep-ph/0406207; ibid, JHEP 0203 (2002) 002.

[18] A. Connes, Academic Press, Inc., San Diego, CA, 1994.

[19] J. Madore, London Mathematical Society Lecture Note Series, Vol. 257, Cambridge University Press, Cambridge, 1999.

[20] M. Buric et al., JHEP 0604 (2006) 054; Buric M., Madore j., Zoupanos G., SIGMA 3:125,2007, [arXiv:0712.4024 [hep-th]].

[21] T. Filk, Phys. Lett. B 376 (1996) 53; J. C. Várilly and J. M. Gracia-Bondía, Int. J. Mod. Phys. A 14 (1999) 1305 [hep-th/9804001]; M. Chaichian, A. Demichev and P. Presnajder, Nucl. Phys. B 567 (2000) 360 [hepth/ 9812180]; S. Minwalla, M. Van Raamsdonk and N. Seiberg, JHEP 0002 (2000) 020 [hep-th/9912072].

[22] H. Grosse and R. Wulkenhaar, Lett. Math. Phys. 71 (2005) 13 [hep-th/0403232].

[23] H. Grosse and H. Steinacker, Adv. Theor. Math. Phys. 12 (2008) 605 [hep-th/0607235]; ibid, Nucl. Phys. B 707 (2005) 145 [hep-th/0407089].

[24] A. Connes, J. Lott, Nuclear Phys. B Proc. Suppl. 18 (1991), 29-47; A.H. Chamseddine, A. Connes, Comm. Math. Phys. 186 (1997), 731-750, hep-th/9606001; A.H. Chamseddine, A. Connes, Phys. Rev. Lett. 99 (2007), 191601, arXiv:0706.3690.

[25] C.P. Martín, M.J. Gracia-Bondía, J.C. Várilly, Phys. Rep. 294 (1998), 363-406, hep-th/9605001.

[26] M. Dubois-Violette, J. Madore, R. Kerner, Phys. Lett. B217 (1989), 485-488; ibid, Classical Quantum Gravity 6 (1989), 1709-1724; ibid, J. Math. Phys. 31 (1990), 323-330.

[27] J. Madore, Phys. Lett. B 305 (1993), 84-89; ibid, (Sobotka Castle, 1992), Fund. Theories Phys., Vol. 52, Kluwer Acad. Publ., Dordrecht, 1993, 285-298. hep-ph/9209226.

[28] A. Connes, M.R. Douglas, A. Schwarz, JHEP (1998), no.2, 003, hep-th/9711162.

[29] N. Seiberg, E. Witten, JHEP (1999), no.9, 032, hep-th/9908142.

[30] N. Ishibashi et al., Nucl. Phys. B498 (1997) 467 [arXiv:hep-th/9612115].

[31] B. Jurčo et al., Eur. Phys. J. C 17 (2000), 521-526, hep-th/0006246; B. Jurčo, P. Schupp, J. Wess, Nuclear Phys. B 604 (2001), 148-180, hep-th/0102129; B. Jurčo, et al., Eur. Phys. J. C 21 (2001), 383-388, hep-th/0104153; G. Barnich, F. Brandt, M. Grigoriev, JHEP (2002), no.8, 023, hep-th/0206003.

[32] M. Chaichian et al., Eur. Phys. J. C 29 (2003), 413-432, hep-th/0107055.

[33] X. Calmet et al., Eur. Phys. J. C 23 (2002), 363-376, hep-ph/0111115; P. Aschieri et al., Nuclear Phys. B 651 (2003), 45-70, hep-th/0205214; W. Behr et al., Eur.Phys.J.C29: 441-446, 2003.

[34] P. Aschieri et al., JHEP (2004), no. 4, 034, hep-th/0310072; ibid, Fortschr. Phys. 52 (2004), 718-723, hep-th/0401200; Aschieri P. et al., Conference: C04-08-20.1 (2005) 135-146, hep-th/0503039.

[35] P. Aschieri et al., JHEP (2006), no. 9, 026, hep-th/0606021; P.Aschieri et al., Apr 2007, arXiv:0704.2880.

[36] H. Steinacker, G. Zoupanos, JHEP (2007), no. 9, 017, arXiv:0706.0398.

[37] A. Chatzistavrakidis, H. Steinacker and G. Zoupanos, Sep 2009. Published in Fortsch.Phys. 58 (2010) 537-552 arXiv:0909.5559 [hep-th].

[38] A. Chatzistavrakidis, H. Steinacker and G. Zoupanos, JHEP 1005 (2010) 100 [arXiv:hep-th/1002.2606]; A. Chatzistavrakidis and G. Zoupanos, SIGMA 6 (2010) 063 [arXiv:hep-th/1008.2049].

[39] D. Lüst and G. Zoupanos, Phys. Lett. B165 (1985) 309; G. Douzas, T. Grammatikopoulos and G. Zoupanos, Eur. Phys. J. C59 (2009) 917

[40] D. Kapetanakis and G. Zoupanos, Phys. Lett. B249, 73(1990); ibid., Z. Phys. C56, 91 (1992).

[41] J. Madore, Classical Quantum Gravity 9 (1992), 69-87.

[42] T. R. Taylor and G. Veneziano, Phys. Lett. B212, 147(1988).

[43] K. R. Dienes, E. Dudas and T. Gherghetta, Nucl. Phys. B537, 47(1999).

[44] T. Kobayashi et al., Nucl. Phys. B550, 99 (1999); J. Kubo, H. Terao and G. Zoupanos, Nucl. Phys. B574, 495(2000); ibid., Oct 2000, [arXiv:hep-ph/0010069].

[45] C. Wetterich, Nucl. Phys. B222, 20 (1983); L. Palla, Z.Phys. C 24, 195 (1984); K. Pilch and A. N. Schellekens, J. Math. Phys. 25, 3455(1984); P. Forgacs, Z. Horvath and L. Palla, Z. Phys. C30, 261(1986); K. J. Barnes et al., Z. Phys. C33, 427(1987).

[46] E. Witten, Phys. Lett. B144, 351(1984).

[47] Y. Hosotani, Phys. Lett. B126, 309(1983); B129, 193(1983); E. Witten, Nucl. Phys. B258, 75(1985); J. D. Breit, B. A. Ovrut and G. C. Segre, Phys. Lett. B158, 33(1985); B. Greene, K. Kirlkin and P. J. Miron, Nucl. Phys. B274, 575(1986); B. Greene et al., Nucl. Phys. B278, 667(1986).

[48] G. Zoupanos, Phys. Lett. B201, 301(1988).

[49] N. Kozimirov, V. A. Kuzmin and I. I. Tkachev, Sov. J. Nucl. Phys. 49, 164(1989); D. Kapetanakis and G. Zoupanos, Phys. Lett. B232, 104(1989).

[50] K.S. Babu, X.-G. He and S. Pakvasa, Phys. Rev. D33 (1986) 763; G.K. Leontaris and J. Rizos, Phys. Lett. B632 (2006) 710; J. Sayre, S. Wiesenfeldt and S. Willenbrock, Phys. Rev. D73 (2006) 035013, [hep-ph/0601040].

[51] E. Ma, M. Mondragon and G. Zoupanos, JHEP 0412 (2004) 026; S. Heinemeyer, E. Ma, M. Mondragon and G. Zoupanos, AIP Conf. Proc. 1200 (2010) 568 [arXiv:0910.0501 [hep-ph]].

[52] S.P. Trivedi, S. Vaidya, JHEP (2000), no. 9, 041, hep-th/0007011; B.P.Dolan, O. Jahn, Internat. J. Modern Phys. A 18 (2003), 1935-1958, hep-th/0111020.

[53] A.P. Balachandran et al., J. Geom. Phys. 43 (2002), 184-204, hep-th/0107099; U. Carow-Watamura, H. Steinacker, S. Watamura, J. Geom. Phys. 54 (2005), 373-399, hep-th/0404130; B.P. Dolan et al., JHEP (2008), no. 3, 029, [arXiv:0711.1347].

[54] J. Madore et al., Eur. Phys. J. C 16 (2000) 161-167, hep-th/0001203.

[55] D. Harland, S. Kurkçuoğlu, Nucl. Phys. B 821 (2009), 380-398, [arXiv:0905.2338].

[56] S. Kachru, E. Silverstein, Phys. Rev. Lett. 80 (1998), 4855-4858, hep-th/9802183.

[57] P. Candelas et al., Nucl. Phys. B 258 (1985), 46-74.

[58] J.P. Gauntlett, D. Martelli, D. Waldram, Phys. Rev. D 69 (2004), 086002, hep-th/0302158.

[59] J.M. Maldacena, Adv. Theor. Math. Phys. 2 (1998), 231-252, hep-th/9711200; ibid, Internat. J. Theoret. Phys. 38 (1999), 1113-1133.

[60] M.R. Douglas, G.W. Moore, hep-th/9603167.

[61] M.R. Douglas, B.R. Greene, D.R. Morrison, Nucl. Phys. B 506 (1997), 84-106, hep-th/9704151.

[62] N. Arkani-Hamed, A. G. Cohen and H. Georgi, Phys. Rev. Lett. 86 (2001) 4757 [hep-th/0104005]; Phys. Lett. B 513 (2001) 232 [arXiv:hep-ph/0105239].

[63] H. Steinacker, Nucl. Phys. B 679 (2004), 66-98, hep-th/0307075.

[64] D. Bailin, A. Love, Phys. Rep. 315 (1999), 285-408.

[65] L.J. Dixon et al., Nuclear Phys. B 261 (1985), 678-686; ibid, Nucl. Phys. B 274 (1986), 285-314.

[66] L. Brink, J.H. Schwarz, J. Scherk, Nucl. Phys. B 121 (1977), 77-92; F. Gliozzi, J. Scherk, D.I. Olive, Nucl. Phys. B 122 (1977), 253-290.

[67] A.E. Lawrence, N. Nekrasov, C. Vafa, Nucl. Phys. B 533 (1998), 199-209, hep-th/9803015.

[68] A. Djouadi, Phys. Rep. 459 (2008), 1-241, hep-ph/0503173.

[69] H. Steinacker, Springer Proceedings in Physics, Vol. 98, Springer, Berlin, 2005, 307-311, hep-th/0409235.

[70] S.L. Glashow, Published in Providence Grand Unif.1984:0088, 88-94.

[71] V.A. Rizov, Bulg. J. Phys. 8 (1981), 461-477.

[72] E. Ma, M. Mondragón, G. Zoupanos, JHEP (2004), no. 12, 026, hep-ph/0407236.

[73] G. Lazarides, C. Panagiotakopoulos, Phys. Lett. B 336 (1994), 190-193, hep-ph/9403317.

[74] S. Heinemeyer, E. Ma, M. Mondragón, G. Zoupanos, AIP Conf. Proc. 1200 (2010) 568-571, arXiv:0910.0501.

[75] K.S. Babu, X.G. He, S. Pakvasa, Phys. Rev. D 33 (1986), 763-772.

[76] G.K. Leontaris, J. Rizos, Phys. Lett. B 632 (2006), 710-716, hep-ph/0510230.

[77] J.E. Kim, Phys. Lett. B 564 (2003), 35-41, hep-th/0301177; K.S. Choi, Kim J.E., Phys. Lett. B 567 (2003), 87-92, hep-ph/0305002.

[78] H. Grosse, F. Lizzi and H. Steinacker, Phys. Rev. D81 (2010) 085034, [arXiv:1001.2703 [hep-th]]; H. Steinacker, Nucl. Phys. B 810 (2009) 1 [arXiv:0806.2032 [hep-th]].

[79] J. Maalampi and M. Roos, Phys. Rept. 186 (1990) 53.

[80] S. Heinemeyer, M. Mondragon and G. Zoupanos, Int.J.Mod.Phys. A29 (2014) 18, [hep-ph/1430032].

[81] M. Mondragon, N. Tracas and G. Zoupanos, [arXiv:1403.7384 [hep-ph]].

[82] S. Heinemeyer, M. Mondragon and G. Zoupanos, SIGMA 6 (2010) 049 [arXiv:1001.0428 [hep-ph]].

[83] A. Chatzistavrakidis, H. Steinacker, , G. Zoupanos, PoS CORFU2011, PoC: C11-09-04.1, arXiv:1204.6498 [hep-th].

[84] A. Chatzistavrakidis, H. Steinacker, G. Zoupanos, JHEP 1109 (2011) 115, arXiv:1107.0265 [hep-th]

[85] T. Banks, W. Fischler, S. H. Shenker and L. Susskind, Phys. Rev. D 55 (1997) 5112 [hep-th/9610043].

[86] I. Chepelev, Y. Makeenko and K. Zarembo, Phys. Lett. B 400 (1997) 43 [hep-th/9701151]; A. Fayyazuddin and D. J. Smith, Mod. Phys. Lett. A 12 (1997) 1447 [hep-th/9701168]; H. Aoki, N. Ishibashi, S. Iso, H. Kawai, Y. Kitazawa and T. Tada, Nucl. Phys. B 565 (2000) 176 [hep-th/9908141].

[87] S. Iso, Y. Kimura, K. Tanaka and K. Wakatsuki, Nucl. Phys. B 604 (2001) 121 [hep-th/0101102]; Y. Kimura, Prog. Theor. Phys. 106 (2001) 445 [hep-th/0103192]; Y. Kitazawa, Nucl. Phys. B 642 (2002) 210 [hep-th/0207115].

[88] H. Aoki, S. Iso and T. Suyama, Nucl. Phys. B 634 (2002) 71 [arXiv:hep-th/0203277].

Talks not Submitted to the Proceedings

Irina Ya. Arefeva
Holographic description of heavy-ions collisions

Aleksandar Miković
Regge quantum theory of gravity

Marko Vojinović
The cosmological constant problem in the spincube model of quantum gravity

Goran Djordjević
"Classicalization" vs quantization of tachyon dynamics

Branko Dragovich
p-Adic mathematical physics and its applications

Sergey Vernov
Exact and asymptotic solutions for cosmological models with non-minimally coupled scalar fields

Ekaterina Pozdeeva
Particle creation multiplicity in modified AdS_5 *spaces*

Dimitri Polyakov
Higher spin gauge theories, background independence and string field theory

Predrag Jovanović
Gravitational lensing as a powerful tool in observational cosmology

Ivan Dimitrijević
On nonlocal modified gravity with cosmological solutions

Takeo Kojima
Vertex operator approach to semi-infinite lattice model

Voja Radovanović
Noncommutative SO(2,3) gauge theory and noncommutative gravity

List of Participants

1. Ciprian Sorin Acatrine (Bucurest, Romania)
2. Irina Aref'eva (Moscow, Russia)
3. Andrey Bagrov (Moscow, Russia)
4. Ilya Bakhmatov (London, UK)
5. Alexander Bakulev (Dubna, Russia)
6. Antun Balaž (Belgrade, Serbia)
7. Xavier Bekaert (Tours, France)
8. Neven Bilić (Zagreb, Croatia)
9. Milutin Blagojević (Belgrade, Serbia)
10. Loriano Bonora (Trieste, Italy)
11. Nikolay Bulatov (Moscow, Russia)
12. Maja Burić (Belgrade, Serbia)
13. Nikola Burić (Belgrade, Serbia)
14. Alexander Burinskii (Moscow, Russia)
15. Martin Cederwall (Goteborg, Sweden)
16. Constanta-Dana Constatinescu (Craiova, Romania)
17. Radu Constatinescu (Craiova, Romania)
18. Branislav Cvetković (Belgrade, Serbia)
19. Ljubica Davidović (Belgrade, Serbia)
20. Ivan Dimitrijević (Belgrade, Serbia)
21. Marija Dimitrijević (Belgrade, Serbia)
22. Goran Djordjević (Niš, Serbia)
23. Veljko Dmitrašinović (Belgrade, Serbia)
24. Vladimir Dobrev (Sofia, Bulgaria)
25. Branko Dragovich (Belgrade, Serbia)
26. Vladimir Dragović (Belgrade, Serbia)

27. Miroljub Dugić (Kragujevac, Serbia)
28. Jelena Grujić (Belgrade, Serbia)
29. Siniša Ignjatović (Banja Luka, Bosnia and Herzegovina)
30. Dejan Joković (Belgrade, Serbia)
31. Božidar Jovanović, (Belgrade, Serbia)
32. Hedvika Kadlecova (Prague, Czech Republic)
33. Mikhail Katanaev (Moscow, Russia)
34. Takeo Kojima (Yonezawa, Japan)
35. Alexey Koshelev (Brussels, Belgium)
36. Neil Lambert (London, UK)
37. Michel Lapidus (Riverside, USA)
38. Marija Marjanović (Belgrade, Serbia)
39. Miodrag Mateljević (Belgrade, Serbia)
40. Aleksandar Miković (Lisbon, Portugal
41. Mikhail Monastyrsky (Moscow, Russia)
42. Stana Nikčević (Belgrade, Serbia)
43. Biljana Nikolić (Belgrade, Serbia)
44. Bojan Nikolić (Belgrade, Serbia)
45. Emil Nissimov (Sofia, Bulgaria)
46. Matej Pavšič (Ljubljana, Slovenia)
47. Dragan Popović (Belgrade, Serbia)
48. Dragan Prekrat (Belgrade, Serbia)
49. Voja Radovanović (Belgrade, Serbia)
50. Zoran Rakić (Belgrade, Serbia)
51. Igor Salom (Belgrade, Serbia)
52. Dragan Savić (Belgrade, Serbia)
53. Branislav Sazdović (Belgrade, Serbia)
54. Djordje Šijački (Belgrade, Serbia)

55. Marko Simonović (Belgrade, Serbia)
56. Silviu-Constantin Sararu (Craiova, Romania)
57. Paul Sorba (Annecy, France)
58. Fumihiko Sugino (Okayama, Japan)
59. Anton Trushechkin (Moscow, Russia)
60. Milovan Vasilić (Belgrade, Serbia)
61. Sergey Vernov (Moscow, Russia)
62. Anca Visinescu (Bucharest, Romania)
63. Mihai Visinescu (Bucharest, Romania)
64. Marko Vojinović (Belgrade, Serbia)
65. Igor Volovich (Moscow, Russia)
66. Alexander Zakharov (Moscow, Russia)
67. Rade Živaljevic (Belgrade, Serbia)